The New Agritourism:
Hosting Community & Tourists on your Farm

Barbara Berst Adams

New World Publishing
Auburn, California

The New Agritourism:
Hosting Community & Tourists on your Farm

By Barbara Berst Adams

Cover painting: "Visiting the Dahlia Fields," by Kipp Davis, Anacortes, Washington

Photos: Unless otherwise noted, all photos by Kipp Davis and Barbara Adams

Publisher's note: We believe this book provides the best possible guidelines available for starting and growing an agritourism enterprise; yet the complex subject of agritourism cannot be mastered from a book. Every farm's bioregion and community are unique, and the art and science of agritourism evolve continually with new techniques and challenges appearing constantly. Therefore you will need to take full responsibility for the specific, how-to decisions for your unique situation and locale. This book provides an abundance of resources for further research beyond the scope of this book.

Disclaimer: This book should be used only as a guide and not as an ultimate source of agritourism information. It is sold with the understanding that the publishers and author are not engaged in rendering legal, health, financial or any other professional services. For this, you should seek a professional about your particular location and situation. Therefore, the author and publisher disclaim any liability, loss risk, personal or otherwise, which is incurred as a consequence, directly or indirectly because of the use and application of the contents of this book.

Publisher's Cataloging-in-Publication
(Provided by Quality Books, Inc.)

Adams, Barbara Berst, 1957-
      The new agritourism : hosting community & tourists on
your farm / Barbara Berst Adams.
         p. cm.
      Includes bibliographical references and index.
      ISBN-13: 978-0-9632814-4-9
      ISBN-10: 0-9632814-4-5

      1. Agritourism. 2. New business enterprises.
   I. Title.

S565.88.A33 2008   338.1'068
                              QBI08-600051

New World Publishing; 11543 Quartz Dr. #1; Auburn, California 95602. www.nwpub.net

Dedication

Dedicated to my many farmer, rancher and horticulturist ancestors, including Amelia Rosen and Dan Dineen.

Acknowledgements

The first person to thank has to be my husband, Kipp Davis, whose loving presence often accompanies me to the many farms I visit. Whether negotiating curious goats with camera in hand, driving us through unknown backroads, or getting me laughing over the many pleasures we come across during our excursions, he's the partner I always dreamed of having by my side through life's adventures.

I want to thank Eric Gibson and New World Publishing for their dedication to farmers and for their attention to detail and quality.

Thanks very much to the research organizations including the Leopold Center for Sustainable Agriculture, and the Rodale Institute for their admirable work towards an eco-friendly and farmer-friendly new society.

In particular at Rodale, thanks to my colleague, Dan Sullivan, for helping spread the advancing knowledge of sustainable and regenerative farming via the written and spoken word, guiding me to various farmers to interview, and for his fun sense of humor.

I am so grateful to the many leaders and groups involved in helping the cause of agritourism, from Nikki Rose in Crete to Jane Eckert in the USA and the many others also quoted in this book who offered information, interview time and photos.

Finally, thanks so much to the farmers themselves who took the time to answer my many questions, allowed us to come to their farms for photos, and sometimes even sent photos to us to help with this cause. You are the heroes and the backbone of a sustainable new world.

Contents

Foreword

Every October my family makes the trek down the hill from our home to Savidge Farms in Alburtis, Pa., where a vast U-pick pumpkin patch is accompanied by a miniature golf course, a corn maze, a combine-turned-jungle-gym and a gift shop. While we grow pumpkins as part of our cover-crop and no-till experiments at the organic research farm where I work—and the farm operations guys are generous with handing them out to employees—we nevertheless make the dedicated trip to the Savidges each year for the haywagon ride, the ritual of picking out and hauling in our own pumpkin from the field, and to support our neighbors.

So when Bear Creek Alpacas opened its barn doors to the public another half-mile down the road a few months ago, we immediately lined up a field trip for our home-school cooperative. Each student came back the proud owner of a "baby alpaca" purchased at the farm's gift shop, and the kids went on for days about the awesome "spit fight" they witnessed between two of furry, feisty critters. (From the parents' slightly loftier perspective, each child's visit to a working farm moulds a more conscious consumer and steward of the land.)

Right here at the 333-acre Rodale Institute, a nonprofit education and research farm, our young and hugely successful CSA farming couple has become a main attraction. What began as a bit of a gamble for John and Aimee Good—who left another thriving CSA they were managing to begin anew

with us—has turned into aces for everyone involved. More than 200 regional customers get to visit the institute once a week to pick up their box of vegetables from the Quiet Creek Farm CSA, perhaps visit the U-pick berry, bean and flower gardens open to CSA members, and walk our perimeter trail or browse our bookstore. The Goods and their customers have brought new energy and vitality to our farm.

Yes, The New Agritourism is alive and well in Berks County, Pennsylvania, and across the United States, as families and individuals wander collectively, eyes squinting, out of the shopping malls and mega superstores and begin the quest for authenticity. Fred Kirschenmann of the Leopold Center for Sustainable Agriculture says these newly awakened eaters are searching for "food with a story." Sure, Americans want something fun to do on Saturday, but they're looking for a whole lot more. They desire to know where there food comes from, who produced it, and that it was grown or raised in a manner that will sustain their bodies and the environment for generations to come.

The New Agritourism offers the growing disenchanted an antidote to the dubious science, animal cruelty, mass marketing and mass transit that's become part and parcel of America's food system (to use the word "food" loosely). It's a movement that puts farmers back into the well-deserved seat of respectability within their communities and gives consumers opportunities to make more informed, healthy choices. The New Agritourism cuts out the middlemen and puts control back into the hands of farmers and those they serve (and more dollars in the farmers' pockets). It's nothing short of a quiet revolution.

And what's a revolution without a manifesto? *The New Agritourism: Hosting Community and Tourists on Your Farm* is a must-read for anyone considering making a go of farming, or for anyone who is not quite making a go of it and might do well to rethink their operation. Barbara Berst Adams offers a thoughtful and detailed roadmap that dispenses both inspiration and nuts-and-bolts advice, including real-life examples of people who have turned their business around financially, created on-farm income opportunities for all family members and discovered that farming can be fun again.

Slow and steady wins the race, the author councils, and if you're not a people person you might want to reconsider opening your farm to the public (as a journalist within the organic and sustainable farming movement for many years, I've met many farmers who discovered that relationships were the best crop worth cultivating once they established a direct connection with their customers). If you *are* ready to open the gates, it's within these pages that you will learn to develop a business plan, research and weigh the focus and details of your operation, navigate the bureaucracy, market your business and get your legal ducks in a row. You'll learn how to plan on-farm events—concerts and outdoor fine-dining are fair game and profitable—and to harness the efforts of your customers, the local media and the growing "good food" movement to promote your business.

My wife and I have long talked about experimenting with cob construction by building a playhouse for our children in the woods behind our home. So I was delighted to discover resources for cob builders within a chapter on B&Bs in a section suggesting alternative guest accommodations. Farmer-chef events, on-farm internships, turning value-added enterprise into value-added enterprise squared by offering on-farm classes—the best part about this read is that great ideas are backed up with excellent resources, inviting the reader to seek further where they are so inspired.

In our world of global warming and big business as usual, it's easy to feel like We the People have no control. One powerful choice we do exercise daily is how and where we spend our food (and fiber) dollars. More and more people are coming to understand this, and they are seeking out opportunities to support real change through this daily act of conscious (versus "conspicuous") consumerism. By embracing The New Agritourism, you become a part of that change, even as you find new and enjoyable ways to improve your bottom line.

– Dan Sullivan, Senior editor, Rodale Institute and former Senior editor at *Organic Gardening* magazine. ❧

Preface

There were two daughters and a son living in the happy farming home of Jane and Jack Hogue in Odebolt, Iowa. But there was a problem the family needed to solve.

In spite of Jack's hard work as a hog and row-crop farmer, there wasn't enough income to support their family of five. Jane, previously a town girl, knew little about farming and gardening until she married and came to live on the farm with Jack. Now with three kids, she had to consider getting a job in town, which would have made it difficult to devote the time she wanted to spend at home with her family. So from that problem, a new idea sprouted: agritourism.

Jane's move to the country as a bride had inspired her to discover a passion she didn't know, until then, would enchant her. She began planting flower seeds and watching them grow into beautiful borders around her country home. Jane says she had missed playing in the dirt as a kid, and that her childhood experience with plants was limited to the Styrofoam cup planted for every Mother's Day during Girl Scout meetings. She wanted something different for her own children, and so set out to cultivate a stimulating environment for them. She called it a sort of outdoor classroom of beautiful flowers, fresh vegetables, mysterious herbs, earthworms, butterflies and living soil, all waiting to be explored and studied. When the need for more income became clear, the gardens helped spark an idea that looked much brighter than working in town.

Jane opened an on-farm business, The Prairie Pedlar. Customers came to purchase nursery plants she grew herself and items made from her other crops, as well as garden gifts from other local farmers and artisans. The visitors were delighted by her surrounding gardens; the business seemed to be thriving, and Jane saw what a great draw the gardens were for gaining farm store customers. When seven acres next door came up for sale, the Hogues bought the property, and expanded their on-farm business with full-fledged agritourism as the main tool for generating customers. Jane began to build gardens in themes, including "On Holy Ground," a biblical garden; "Tyler's Barnyard Garden," a garden of flowers and herbs that have names correlating with farm animals (lamb's ear, hens&chicks, horseradish…); "Kinder Garden," where there's a plant for every letter of the alphabet; and "Moon Garden." "The Moon Garden illuminates the evening as moonlight settles on white patches of flowers," says Jane. "In daylight this garden is a matinee of white refinement, but each night it transforms to an iridescent glow of moonstruck blossoms, setting the stage for mystery and enchantment. The moon flower is the star attraction, with mother-of-pearl, yarrow, white veronica, physostegia (false dragonhead), lamb's ear, baby's breath, Iberian candytuft, and daisies cast in the supporting roles. The darkness is accented with white statuary, but the night-scented flowers like white nicotiana, fragrant stock, and evening primrose steal the show!"

The Hogues now add special events to their agritourism package. "Our Moonlight Garden Party now has come and gone for the 11th year in a row," says Jane. "Some people make reservations months ahead of the party to guarantee a spot. Many who come are wowed by the details of the evening – the lighting, candles, luminaries, good food, music, etc., etc. It's a charming affair and the gardens are magical at night. People often forget to enjoy their flowers by moonlight and we use this evening to promote that. While all of the bright colors of the afternoon garden recede into the shadows of night, all of the white and pastel flowers are illuminated by the moon. The sounds and smells of the night time garden are a bit mysterious as well. As the sun sets on the gardens it takes on a life of its own and the Moonlight Garden Party has always been a favorite event for customers who come back year after year. The Moonlight Garden Party is a great PR event and gives us a chance to 'wine-and-dine' our customers, and many want a memento to remember the evening by, so our gift shop benefits too."

The Hogues also especially enjoy agritourism events with children, seeing as though the gardens themselves were inspired by the desire to bring nature to their own children. "We are in the seventh year of our Green Team program," says Jane. "Children ages 8 – 12 participate at the garden every Wednesday morning from 8:30 to 11:00. This summer 39 students come with eager anticipation, a sense of wonderment and energy that is impossible to contain. We delight in their enthusiasm for nature and hope that we plant a seed that will make them life-long gardeners. Each morning, the children tend to their own garden space (the Secret Garden at the center of the Garden Maze) learning to plant, weed, mulch, water, dead-head, etc. After a light snack, we head to the country schoolhouse on the property for a craft session so that they can take

> "When we started our business 22 years ago," says Jane, "I received some very good advice from another entrepreneur. She said that I should not depend on local customers to keep my business going. She encouraged us to reach way beyond our local boundaries and promote our gardens as a destination spot in the country."
>
> — Jane Hogue, The Prairie Pedlar, Odebolt, IA

home a completed project each week. This program receives excellent publicity and goes a long way in enhancing our public relations efforts in the area. It is also a retail opportunity because parents bring their children to the garden and browse around, shop in the gift shop, etc."

The Prairie Pedlar now attracts customers from far and wide. Up to 5,000 guests show up annually and sell-out crowds enjoy their special events. "When we started our business 22 years ago," says Jane, "I received some very good advice from another entrepreneur. She said that I should not depend on local customers to keep my business going. She encouraged us to reach way beyond our local boundaries and promote our gardens as a destination spot in the country." Often, the gardens serve as a lure to draw customers to the farm store who spend money once they are there. Other times, the gardens themselves provide income in the form of fees charged for special events, or rental of the property for private gatherings such as bridal showers, weddings and family reunions.

Jack blends his original farming operation with the Hogue agritourism well. "Gardening and farming are very similar occupations," says Jane, "and many of the same tasks occur at the same time (planting, weeding/cultivating, harvest, etc.), so he is a busy person. Jack manages the greenhouses and

At Prairie Pedlar, more than seventy-five display gardens are neatly arranged into charming theme beds that attract visitors throughout the summer. Farm visitors then often stop off at the greenhouses and gift shop where the farmers' annuals, rare perennials, scented geraniums, herbs, and other items including value-added products from field-grown flowers are for sale.

Photos courtesy Prairie Pedlar. Jane Hogue, photographer.

Mother and kids enjoying U-pick lavender field on agritourism farm.

plant production from March - May. He also enjoys the garden visitors and is host for the many garden tours that we have each summer. He leads the guided tours, sharing much plant lore and flower stories that charm visitors."

Today, there are two grown daughters and a grown son, now along with their own spouses, who get to come home to visit the farm life that was allowed to continue because of an agritourism project that sprouted many years ago. Jane and Jack's youngest daughter was their final child to marry, and recently wed on the farm to her high school sweetheart. This inspired the Hogues to add even more customer attractions to their rural business. "We've added a new wedding chapel and a terrace garden with fountain for the occasion. We moved an old ten-sided granary onto the property last fall and have transformed it into sort of an old fashioned bandstand like were in many city parks in the Midwest. It will make a wonderful setting for the ceremony."

The Hogue's agritourism began as a family project to benefit their own kids, and continues to be even as their children grow up. As with many of the most successful agritourism operations, the Prairie Pedlar's crops and agritourism are extensions of a family's ever-changing, preferred rural way of living that adds to the family's pleasure, rather than becoming an added burden that dilutes their chosen farming lifestyle. "It is a very rewarding opportunity for us to work closely with kids and try to instill an appreciation for flowers, herbs, and the rewards of gardening," says Jane. "They delight in the simplest things and manage to find garden surprises that we sometimes miss. Enjoying the gardens through the eyes of an energetic ten-year-old keeps us young at heart and reminds us on a weekly basis of all the reasons we love to garden." ✿

Part I: Introduction

Agritourism: Is it really new? Can you benefit from it today?

1

Several years ago, the United States Department of Agriculture reported that an Indiana sustainable farmer invited 50 chefs to his farm during a one-time summer "expo," which resulted in restaurants listing his farm on their menus, purchasing his products as promotions, and an estimated tripling of his restaurant sales. This may have been a surprise, certainly a pleasant one, for any farmer in North America back then to discover that inviting non-farming citizens onto the farm could make such a positive impact.

But today, farmers around the world may want to take note. Very recently, the National Geographic Society director of sustainable tourism, Jonathan Tourtellot, told a group of journalists we should prepare for something brand new… tourism on steroids. Based partly on the near doubling of international tourism as the last century turned, as we reach 2015, the tourist forecast for those traveling beyond their homeland was reported at a billion and a half. When projections for the people expected to take vacations close to home is added in, within the next decade the estimate is seven billion tourists exploring their neighborhoods and faraway lands. And the good news is that there's a growing push by numerous groups, both regional and international, to make sure this new tourism is the kind that helps local cultures and economies sustain and enhance

their uniqueness. The travelers also support this with their growing desire to leave behind generic resorts for authentic, responsible eco-tourism, of which agritourism is an important branch. Already, in the U.S.A. alone, the U.S. Forest Service reported 62 million Americans took part in some form of agritourism, from day visits to overnight stays to extended educational stays.

Jane Eckert, founder of Eckert AgriMarketing, reports how agritourism—hosting non-farming community and tourists on the farm in a multitude of ways—can bring remarkable benefits to farmers just in the month of October alone: "The October season has become big business for many farms that want to venture into agritourism. Depending on the farm location and ability to attract customers, farms not only sell pumpkins, but food concessions, fall decorations, make-your-own scarecrows, wagon rides, corn mazes and other family activities. Many fall season farms are actually charging a general admission just to come onto the farm.

"While the average pumpkin sale might be $4 – $8 per customer," Eckert continues, "they will generally spend at least $20 per family just to have a fun day on the farm. Fall season revenues might start for farms at just a few thousand dollars. But with a little bit of ingenuity, hard work and a good product mix,

$100,000 is not a difficult goal for a farm to reach in October. After several years, many farms are approaching sales up to $500,000 and more.

"Most farms I know exceed $100,000 annually from their October season. The concept is start small with pumpkins and then start adding the products, food sales, school tours, etc., and the revenue builds quickly."

Agritourism, of course, ranges from cut-flower U-picks to on-farm, artisan cheese-making classes to farm B&Bs. The possibilities are endless. But so, it seems, are the potential agritourism customers.

Studies have shown that millions of non-farming citizens yearn to make direct contact with farming and rural lifestyles beyond just driving past them on the highway or reading Farmer John storybooks to their children. Long ago in North America and most likely other countries and continents, if we weren't one of the nearly half of the citizens who were farmers, our cousins or grandparents were, and we took Sunday drives out to visit them. There's no place for non-farming folk to go home to now.

Farmers worldwide are affected by this and are tapping into it, with Europe having a headstart on agritourism over many countries. My husband's and my recent visit to France's famous countryside included overnight stays at an old stone farmhouse and working dairy in the heart of the fertile French Louvre Valley operated by a generational farm family. We also spent a night at an historic cider house in France's Normandy apple country, and walked the green fields and walnut groves while staying on a farm in the Dordogne-Perigord area operated since many generations back. Recently in the United Kingdom, one estimate of a network of farms that connects with overnight guests states it puts nearly $73 million in the pockets of farmers. And that's just the segment of agritourism engaged in overnight stays. Nigel Embry, chief executive of Farm Stay, UK, states that his organization's estimate for overall agritourism for the UK would be closer to $96 million. Certainly, some European countries have

subsidy programs that help farmers build agritourism structures that other countries don't enjoy. It may be a program others can learn from.

But, as shown by the successful USA farms described by Jane Eckert and others, subsidies aren't imperative for agritourism to help farmers succeed and need not be depended on. To illustrate further, forget the dollar amounts in the millions for a moment. A small U-pick cut flower garden brings in extra spending money for a single mother in Canada so her daughter can attend summer camp. And a shack town built on an old garbage dump, with no running water, no electricity, no sewage in Cape Town, South Africa, is considered home to the poorest of the poor. A non-profit group there, a "sister" to the author's own non-profit, is helping the country lift itself out of poverty one small area at a time by putting in a sustainable mini-farm and offering eco-tourism at the site. When our non-profit sent a mere $10 worth of open-pollinated seeds to our sister group there, they gratefully organized and shared out every seed to grow them out in gardens for proliferation. Soon, more seeds came and were grown out, and with the blessings of Nelson Mandela, they began an open-pollinated seed bank and started a program for teaching emerging new eco-farmers. From these origins, an entire eco-village funded partially by agritourism, with composting and solar power is emerging from the midst of despair. "For every 200 paying guests, another real cottage can be built for a family living in a make-shift shack," a recent letter from Di Womersley, the director, stated. "The eco-cottages are made from locally sustainable materials, with composting toilets, wind power and organic gardens, and built with help from the people themselves, including unemployed youth. This shack town is slowly being transformed from within to a self-sustaining eco-village."

Though agritourism is still considered in its infancy throughout South Africa, revenue from increased nature-based tourism there is catching on and overlapping with more community-friendly and sustainable eco-tourism. Thanks again in part to Nelson Mandella, local citizens will benefit both from the operation and take part in decision-making. Agritourism in Africa includes thatched-roof accommodations on small farms and tours of fair-trade, rooibos tea plantations, aloe extracting plants and organic banana farms.

Years earlier, in Asia, a group of Japanese women invented Teikei clubs, another specialized form of agritourism which spread to Europe in the mid-60s, then came to the USA in the mid-80s, and became the now classic community supported agriculture–CSA–farm, often a very beneficial form of farmer–non-farming citizen interaction. Agritourism is spreading across the planet from many points of initiation. Subsidies, grants, and other support may help, but the main ingredient for success seems to be the enthusiasm and decision-making ability of those involved, the farmers and their local communities.

A mother enjoys strolling her infant through a special U-pick field grown on a destination lavender farm.

Agritourism's growing popularity can cause problems if foresight isn't used as we'll learn from Nikki Rose, founder of Crete's Culinary Sanctuaries, an internationally acclaimed eco-agritourism network. Nikki is among those concerned that agritourism could be exploited or eventually over-regulated by big industry. "What concerns us," Nikki says, "is that there could be a frightening new trend in 'Disney-agritourism.' Hope you don't mind my candidness, but look what happened to 'eco-tourism.' Shoot, the word is probably already copyrighted by (Big Entertainment Business!)" She is concerned about enticements to get authentic local farms to hook up with big entertainment businesses or large tour companies who insist on cheaper and cheaper prices paid to the locals and farms, pushing authentic local citizens and business into bankruptcy or selling out to become a fabricated attraction. Nikki arranges eco-agritourism programs for people to experience authentic Cretan culture and farming, including organic olive oil production and cheese making. In Crete, sustainable practices have remained intact for thousands of years and can remain so as long any encroaching big business works to their mutual benefit and doesn't try to exploit the "unworldly" locals." Thankfully, she is not alone in her concern about agritourism needing to benefit the authentic farm and local community. "I was on a panel (at the Adventures in Travel Expos in San Francisco) with people who had flashy videos about Belize, Kenya, New Zealand. My presentation focused on meeting the people of Crete, including 'George the Farmer.' I was really intimidated, but pleasantly surprised that the audience was so responsive and referred to our project as 'unique.' They went so far as to ask the other panelists, 'Hey, what are you doing to introduce visitors to locals or support those communities?' There was a lot of fidgeting going on!"

Jane Eckert also points out in a variety of ways that farmers themselves are accountable for becoming and remaining part of the decision making when governments or other large entities oversee farm tourism. It appears there are others around the globe in complete agreement with Nikki and Jane, that for agritourism to succeed, real farmers must help make the decisions, and it must thrive by helping real farms succeed.

"To run an agriturismo," explains Raffaella and Gianluca Zanetta, owners of La Capuccina Farm in Italy, "You must have a working farm and a very professional farmer or manager with in-depth knowledge both of cultivation and agritourism, because agriturismo requires first farming, and second, tourism (both of which are very much connected.)" La Capuccina produces, among other things, milk from their dairy cow, Camilla, which is transformed into butter, yogurt, cream and cheeses. Their goat provides milk for goat cheese. They harvest hazelnuts, walnuts, chestnuts and grow about 50 fruit trees. Their gardens and fields produce onions, potatoes, lavender, and food for their own farm animals. They keep 40 beehives along the farm's edge which bor-

ders the local river Agogna from which they produce what they call Mecalfa and Acacia honey. A donkey and two horses are considered members of their family whom they insist will enjoy living all the way until old age on the family farm.

Agritourism—Is it really something new?

The La Capuccina Farm owners mention (above) that farming and agritourism are very much connected. That might spring from the fact that agritourism, or "non-farming community interacting with farmers for mutual benefit," is a natural social interaction as old as agriculture itself.

Author Donald J. Berg is quoted here from his book, *American Country Building Design,* about what happened in America during the latter half of the 19th century: "Almost every farm was also an inn… Letting rooms was a source of revenue, entertainment and education. Conversations with travelers brought news and new ideas. Immigrants, staying each night at a different farm on their westward trek, would be acclimated to American meth-

Agritourists enjoy a lavender farm.

CHAPTER ONE

A shepherd in the Lassithi Mountains in Crete. 99.9 percent of all cheese made in Crete is from sheep or goat milk, with the animals free to graze wherever they please if not near a metro area. Tourists who embark on programs such as Crete's Culinary Sanctuaries can witness authentic agriculture in a manner that helps sustain local tradition.
Photo by Nikki Rose

ods and designs before they built their own farmstead. It also was common for farm sons to raft a year's harvest down the nearest tributary of the Mississippi to New Orleans, or through the new canals to Philadelphia or New York. The long walk back, with overnight visits to countless farms, was essential to the young farmers' education. An extra bedroom was rarely empty."

Entire villages of long ago once revolved around and celebrated the harvests. Many non-farming French had purple feet during the grape harvest; gatherings abounded during New England maple sugaring time; and Native Americans held planting feasts and ceremonies. From both sides of the equator, from temperate to tropical climates, the draw to keep the authentic farmer connected to the general population is as strong as humankind's urge to gather together in civilized communities. We want to expand, evolve, build great cities and solve universal mysteries, but we also want our farmers by our side (after all, they have expanded, evolved, fed the cities and solved universal mysteries right along with us). As Joseph Bean, a consultant to various Hawaiian agriculture entities, describes about Hawaiian

agritourism: "Over the past decade, Maui's guests have been spilling out of their resorts in growing numbers to enjoy the fields and farms and the forests that have entranced all comers since Hawaii Loa landed here, perhaps 2,000 years ago. Driving through the island's sugar and pineapple fields, walking the taro fields and writing home about the glories of Maui's agriculture, modern tourists are really doing two very interesting things. First," Joseph continues, "they are discovering and participating, if very informally, in the new tourism: agritourism. Second, they are returning Maui tourism to its oldest roots, its actual origin, where Hawaiian values are the rule and real Hawaiian lives are the attraction."

One could almost argue that "real farms" are agritourism farms by default in one form or another, depending on the era and social structure in which they operate, and societies that kept most of their farms isolated from the rest of the culture were occasional experiments, the exceptions. There is security and more progress when local citizens have a sense of ownership of their own food supply, and when farmers and non-farmers understand and appreciate each other.

What about the farmers themselves who engage in agritourism? What's the lifestyle like for them? Many find that hosting community and tourists allows them the satisfaction of telling their story to others, to be heard, understood and appreciated, and to learn likewise of other people's lives and livelihoods. It isn't for every farmer, and doesn't need to be.

Chris Grant is one farmer who knew it was something he wanted to incorporate. Seeing the need and opportunity of agritourism, on a crisp October weekend in rural New York, close to 1,000 people visited the then very new Indian Chimney Farm. This was its grand opening as an agritourism destination. The farm, owned by husband-and-wife team Chris and Kim Grant, totals 65 acres: 35 in pastures and fields, a 25-acre woodlot, and five acres containing two gorges, their home and gardens. In addition to raising and selling alpacas and their wool, they raise honeybees, about 30 free-range hens and

Guinea fowl, offer sustainably harvested firewood and milled Eastern Red Cedar from their forest, and train and sell performance horses. They sell online and through their on-farm store, never taking products off the farm. On this one weekend, nearly 1,000 people connected on the "experience" level with the farm, associating its products and services with a deeper link to good memories, new adventures, and the earth, buying farm products at retail prices direct from the farm, and telling their friends. Research shows that word-of-mouth is more powerful than advertising because a single buyer influences many others due to the trust already well established in social networks. This phenomenon, called "endogenous growth," occurs slowly, but sinks into the network deeply. Over a period of two days, the farm owners learned directly and instantly, what draws their customers to them. "It's funny," says Chris Grant, owner of Indian Chimney Farm along with his wife, Kim Grant, "That we and other adults think the alpaca are a big draw; but to kids, it's the friendly horses that are the best part about visiting

Pumpkin patches are becoming popular fall traditions across the country.

> *The draw to keep the authentic farmer connected to the general population is as strong as humankind's urge to gather together in civilized communities. We want to expand, evolve, build great cities and solve universal mysteries, but we also want our farmers by our side.*

the farm." The networking opportunities blossomed, also. "We have offers from other small businesses to partner in various ways." Indian Chimney Farm had officially plugged itself into the idea of farm visits for promotion, direct sales, and pleasure.

Agritourism can include little or no interaction with other people, such as with an honor system flower bouquet roadside stand, or putting up crop signs for passersby, leading tourists to enjoy the area more, and hopefully spend more time and money purchasing the locality's goods. Agritourism can mean opening up to the public only once per year, such as with an annual sunflower festival or "Apple Harvestfest." It could mean catering to non-farmers on a regular basis, with the farmer choosing specific groups rather than the general public, such as giving tours only to elementary students or their own church members. Some farms include overnight stays in the form of B&Bs, on-farm camping, or by providing country locations for weddings, reunions, or artist or spiritual retreats. While all agritourism is educational, some farmers concentrate more on the serious, educational end, developing farm-to-plate cooking schools for chefs, natural science experiences for preschoolers, handspinning workshops on their wool farm, or bringing on aspiring future farmer interns.

Quillisascut Farm along Washington State's Columbia River is a goat cheese farm where owners Rick and Lora Lea Misterly have farmed since 1987, with Lora Lea having grown up in agriculture. The slogan for their Quillisascut Farm Cheese is "Cheese

from the Pampered Pets of Pleasant Valley." Here on their 36 acres of dairy pasture, wildlands, gardens, orchard and vineyard, along with other nearby networking farms, they established a professional on-farm cooking school.

One chef student who attended the Quillisascut Farm's chef school, Dylan Stockman, has been a professional cook for more than a decade and now works in the Seattle, Washington, USA area. He specializes in cooking Pacific Northwest cuisine, although his establishment draws from all around the world. "I believe that cooks and chefs have so much to learn from opportunities like the farm tour," he says. "When you harvest your own product you have a sense of pride, because YOU harvested it and you get excited about what you are going to do with it. You don't start to think about 'gourmet;' you just want to do it justice; you want the product to stand out, so people can say, 'Wow, I have never tasted a pepper so rich and sweet, where did you get these?' and then you are able to tell them. Then the customers get excited because they can read your excitement." Dylan says that getting one's hands right into the earth, into the soil, creates a profound shift in how one perceives food. "If all the cooks in the world could get a sense of that feeling just once in their lives I honestly believe the kitchen would be a better place. If the cooks could get to the farms or even the farmers markets (which is a religion for our family) and talk to the farmers themselves, it would help their kitchens ten-fold."

Agritourism is attracting farmers both old and new. The Grants of Indian Chimney Farm were relatively new farmers when they first opened their farm gates to agritourism. But Lattin Farms described below is a fifth generation farm that recently decided to take on agritourism by storm by developing programs for school children and offering a vibrant autumn festival. And the owners of Sweet Grass Dairy in Georgia had been dairy farmers for 25 years. After switching from conventional dairying on concrete to sustainable pasture-fed cows and

Farmer Nate, owner of Frog Song Farm, fascinates a group of visitors with his description of his organic farming methods.

goats, they now offer dairy tours with cheese tasting by appointment for groups of 20 or more, twice-a-year open houses, and cheesemaking classes. The venture is fraught with possibility as well as pitfalls. At this point, liability insurance, and some farmers' habit of underselling their products and services, seem to be the major hurdles, or even the insurmountable mountains, that stop some farmers from succeeding. But those hurdles are conquered by a growing number of successful agritourism destinations.

Even aspiring farmers have agritourism in mind. A woman in West Virginia has grown an organic vegetable garden for 25 years, and she and her husband recently bought 18 acres to work towards a farm business. She has no specific agritourism or community visiting plans yet, but she expressed her hopes to the author after reading a previous title, *Micro Eco-Farming*. "I want a piece of land that people want to come to," she says, "to see how to do these things in their own yards and gardens, as well as buying vegetables grown without chemicals on sound, healthy ground."

Overall, as long as agritourism is a product of authentic farms rather than fabricated entertainment, it offers the potential to increase farm revenue, generate public support for local farms, support strong, local economies and help restore the ecosystem. This isn't to say, though, that just anyone should participate in agritourism, or in fact, any farming venture they find unnerving. If you don't like growing mixed vegetables, grow what you understand and like to grow. Be careful if you don't understand or want to learn about groups and crowds as carefully as you would learn the cultivating needs of a new crop, or would prefer to work without "people interruptions."

Some farmers, of course, don't mind people, it's just "visiting" people. Or at least "certain visiting" people.

"The last thing I need is a bunch of slickers parking on the hay field, trying to milk the horses and ride the cows and asking how many eggs my rooster lays."

If people are not part of the picture for you, don't do agritourism.

Agritourism visitors enjoy tours of a farm's lush gardens.

"The next thing will be somebody asking if I have any brown-colored dairy goats where the milk comes out chocolate."

Some, though, have found people to be their favorite crop. And sometimes, forms of agritourism have surfaced within farming families as the hidden sideline or career desire for at least one family member or partner, such as the one who loves to host, meet people from faraway, put on a great party, or gourmet cook for ever-changing audiences. On the generational family Lattin Farms, more than one family member has a love for children and education in teaching, and they bring this love and knowledge onto the farm to create meaningful and educational farm tours, enhance their lifestyle, and strengthen their income. Farmers are experts at weeding and culling what the farm doesn't need, while saving seed and expanding on what works. In many ways, they are naturals not only at agritourism, but being the ones (rather than Big Entertainment Business) to take it from its infancy to higher ground. It's my hope that this book will help you in some way make agritourism a venture as fruitful as all farms deserve to be. ❧

Exploring the Options: 2
A treasury of agritourism possibilities

What can you provide that's unique to your farm that you already do or already have? This chapter provides a treasury of agritourism possibilities for your farm, and also provides insights into leveraging regional assets already available from your community and bioregion. In addition, it is the intention of this chapter to help farmers eliminate overwhelm and choose which type of agritourism activity to try.

Urban life and high-tech lifestyles are some of the agritourism farmer's greatest friends. The "high-tech, high-touch" balance is a phenomenon described by those who study the fluctuations and trends of humankind. The more humans become removed from nature, tangible hands-on creation, and their natural food supply, the more they crave a reconnection to them. What we farmers do everyday has become a variety of mysteries to others, and humans can't get enough mystery: novelty, lost traditions, heirlooms and vintage crafts, new discoveries in earth sciences, post-modern regenerative agriculture, reconnection to nature, and so on.

This chapter explores options for adding agritourism activities to the farm or enhancing current ones. As you will find throughout this book, agritourism usually pays off and allows a good lifestyle for the farmer when the tourism project is an extension of the farm itself, or even if it's quite different, at least it is a complementary secondary passion of at least one of the farm owners. For example, on HeartSong Farm Healing Herbs in New Hampshire, husband and wife Michael and Nancy Phillips, co-owners, have developed a series of on-farm herb classes that grew out of their passions for knowing the herbs while growing wild or on the farm, and the many healing abilities of herbs when they're harvested in a gentle and conscious manner. Nancy now enjoys teaching her passion to others (for more detail about HeartSong Farm, see Chapter 13).

Agritourism isn't new to everyone. Some ventures have been in operation for years. They are the traditional U-picks or on-farm roadside stands. Others may go just one step further and offer a yearly open house of the farm operation. For those exploring even more possibilities, this chapter is a digest of agritourism projects used by farmers across the globe, and may help you come up with your own custom agritourism projects, the one idea (or combination of ideas) that sets your venture in motion. If sales at your roadside stand seem slow, perhaps you'll decide to offer a one-time Saturday walking tour of your fields or a one-time Saturday workshop for children to make flower ornaments set up next to your roadside stand, and contact the media about this time-sensitive event to generate free promotion. You may be looking for ways to add appeal to your CSA on-farm pick-up days to encourage more pick-ups on the farm rather than drop-off sites. Or perhaps you're looking for ways to spice up your current farm tours for school children to attract more teach-

Choosing Agritourism Activities: Don't forget the bottom

To help you further choose activities, determine just how agritourism needs to benefit you. It must bring you more profit in one of three manners:

- Selling more farmed products (hopefully directly) that has come from your efforts in attracting more customers to buy your products. This assumes you have more to sell and need more customers.

- Creating a higher value for your products and getting higher paying markets for your products This assumes you may not have that much more to sell, but just need a fairer, higher, price for what you do have. By attaching tradition and your deeper story to the product in showing non-farmers your operation, you've raised its value and may find a premium market.

- Or charging for the agritourism service itself, such as cooking classes or tours to school children.

If you're already easily selling as much as you can produce, and your government or other situation won't allow higher prices for the foods you produce, then attracting more buyers or adding value to them by attaching your story via agritourism will do you no good and waste your time. Instead, will you be able to charge your own price for actual agritourism services such as tours and classes? If so, that's the type of agritourism to aim for. For this type of agritourism, avoid situations where you sign on with a larger "promotional" entity which will then force you to charge what they dictate for workshops, farm B&B stays, and the like.

If none of the three above are in place, then the agritourism will be a moment of fun for you (a party, a favor to the community), or a charity on your part. Hopefully you've established how much expensive fun, volunteer time, and charity you can give away while replenishing your needs and upgrading your own living conditions for yourself and future farmers. Naturally, if possible, your first ventures will be practice ones that seem more like a party than a business. The future goal is for the upliftment of the business of farming and the revenue farmers receive.

To see more actual agritourism operations and their prices, as well as links to forums with other experienced agritourism enterprises, visit:

www.newagritourism.com.

ers to bring their classes. Here, you'll find activities that can last from one hour to a long weekend or more, and even those that are a regular, weekly happening. And these are only a glance at the iceberg's tip! Listing them all would take volumes, and new ones arise almost daily.

Try not to be overwhelmed by all the possibilities, nor to slip into the thought that you must provide endless hours of blissful fun to an eye-rolling, cross-armed "entertain me," crowd. Continually tweak your approach; weed out the unresponsive audience members and encourage the appreciative ones, just as you would cull your crops and keep only the seeds

of the sweetest melons. In many cases, agritourism income will be aimed at high-end, select audiences rather than low-end, high-volume. Most of those who succeed in agritourism encourage avoiding huge quantities of visitors for no- or low-pay, and instead choose a few one-of-a-kind programs for those who can pay for them. You may find exceptions, such as a free farm tour that pays off well enough with the farm promotion and product sales it generates. But even in that case, the audience must be select and those willing to pay for your products. School children love to encourage their parents to buy from the farm they visited. And school districts

These agritourists enjoy Pelindaba Lavender Farm's specially planted U-pick patch.

hold bake sales or other learning activities to pay for field trips, with the kids loving the activity of earning their own money this way. Mountain bikers, high techies craving the country, and bird watchers are quite financially secure and many are happy to visit your unique farm experience. If you still feel your job is to serve the poor, you can always tithe. For every 20 paying agritourism customers, offer a "scholarship" to a person or family in need. This system makes sure our farmers are well-paid while the disadvantaged are also taken care of.

A Treasury of Farmstay, Workshop, Demonstration, & Festival Themes

* *Fruit (or mint, potato, lavender, etc.) harvesting stays.* Guests become farm hands or even members of a farm family for a weekend, or perhaps a week, and harvest the crops. They are paying for the experience, not the product (as with pumpkin patches). Make sure the labor you're receiving and being paid for gives them some bonuses, as the harvest farmstay is a different than internships, where more serious labor is expected in trade for more serious farm training. Make sure they finish something tangible to give them the satisfaction of completion, such as picking apples or lavender and then seeing it pressed into cider or distilled into essential oil. Give them one of the finished products they helped harvest, if possible: a bag of heirloom potatoes, a bottle of essential oil, a gallon of cider. Before they leave, give them a few photos of themselves on the job, with an attractively illuminated list of the activities they successfully participated in, signed by the farm or ranch owner, and a thank-you card.

* *Mother's or Father's Day tour.* A lilac farm and rhododendron nursery both open their blooming display gardens (which look more like blooming forests) each Mother's Day. These events have become regular Mother's Day events for some families for generations. The farm owners invite a local harpist to play amidst the blossoms, and set up particularly good areas for family photos. A small herb farm offers a farm-owner guided herbal walk, then ends with a session on herbal cooking or herbal kitchen cosmetics, with free herb samples of the farm's value-added herb products for mothers who came with their families as paying guests.

* *Gardening or farming classes* on something mysterious or out-of-the-ordinary, such as astrological

CHAPTER TWO

gardening and farming (for more information, see *Astrological Gardening, Ancient Wisdom of Successful Planting and Harvesting by the Stars* by Louise Riotte.)

- *Winter country holiday workshops:* Make wreaths, garlands or tree ornaments from your farm's evergreens.

- *Home spa and cosmetic recipes and demonstrations* with ingredients from the farm's crops. Promote it at least three weeks before Mother's Day and offer a gift certificate for this class for Mom to come later on her own or with friends.

- *Herbal scented candle-making workshops.*

- *Workshops for nature prints* made from herbs, leaves and flowers collected from the farm.

- *Natural plant dying: T-shirts, raffia, corn husks, and materials for handmade paper.*

- *Flower pressing* for note cards and bookmarks (sometimes onto handmade paper).

- *Old fashioned herbal tea party* revolving around particularly attractive seasons on the farm, and tea herbs (or more precisely, "infusion" herbs) you grow, and/or an historic theme that especially suits your farm. Themes can range from a children's fairy tale tea introducing kids to your farm's chamomile—which Peter Rabbit was given when coming down ill after escaping Mr. McGregor's garden—to a June bridal tea party demonstrating how to make bridal lavender sachets, to a hiker's tea with farm ingredients for men and women who prefer something with a lot less Victorian lace after their invigorating time in your nearby woods. Both men and women appreciate newly discovered health beverages, and your farm can offer its own home-grown combinations. Pelindaba Lavender Farm on San Juan Island in the Pacific Northwest, for example, blends its own lavender with health-giving rooibos tea from South Africa.

- *Food preservation lessons* on drying, freezing, canning. This can be especially good for a farm's

Try not to be overwhelmed by all the possibilities, nor to slip into the thought that you must provide endless hours of blissful fun to an eye-rolling, cross-armed "entertain me," crowd. Continually tweak your approach; weed out the unresponsive audience members and encourage the appreciative ones, just as you would cull your crops and keep only the seeds of the sweetest melons.

own CSA members, or perhaps offered in conjunction with your local agricultural extension.

- *Nature photography workshop:* Invite a local photography buff from a nearby photography store (helps promote his/her store) to give tips on good nature photography, and then allow the students to take photos on the farm.

- *Picnic cooking workshop:* Have students bring their own picnic, minus a main picnic dish that will be demonstrated and prepared from farm-fresh ingredients, then allow them to consume their picnic on the farm.

- *Giftmaking on the farm:* Herbal vinegars, cordials, floral or herb-scented bubblebath, even herb-scented play-dough.

- *Crop-specific celebrations with demonstrations.* For example, a sunflower, mint or lavender festival with demonstrations on cooking with mint or lavender or floral arrangement with sunflowers, making bird feeders from your sunflowers, healing with mint or lavender (invite an aromatherapist as a vendor), and crafting mint sachets and lavender wands. Apple orchardists offer apple pressings, U-picks, apple baking demonstrations and storytelling by Johnnie Appleseed himself. The title, *An Apple a Day,* by Jennifer Storey Gillis offers ideas for kids' games, recipes and apple facts revolving around apples. Heirloom

Growth and Pricing

Pricing happens in stages, with choices made during the planning, and tweaked over time. More precise pricing is mapped out in detail during your formal business planning stage (see Chapter 3), and will be continually evaluated, tweaked, and compared to emerging new numbers such as business expenses and cost of living fluctuations during your agritourism career. This chapter is written just to spark your imagination with possibilities for new or improved ventures. But pricing is mentioned briefly here, also, because sometimes just an initial gross revenue estimate can help determine upfront whether or not you should even take the venture into the business planning stage at all.

Start by determining what local citizens or travelers to your area are paying for activities similar to yours:

- Overnight hospitality stays (look at the local B&B prices);
- Themed weekend or longer retreats (e.g., for writers, artists, and healing retreats);

- Shorter offerings such as two-hour demonstrations or half-day workshops such as art and other hobby or educational workshops (check your local newspaper both to find listings and also to see what facilitators charge);
- Tours of interesting locations not funded by or usually open to the public such as private home and garden tours;
- Facility rentals: See what non-public funded historic buildings, hotels, or display gardens charge for renting as wedding locations or family reunions.
- If their prices seem far too low or high for what you think you could charge or profit from, look closer at what's included in their price. If they're offering weekend retreats far cheaper than you can afford, are they providing the overnight accommodations and food themselves, as you had planned? Participants at events lasting one-day or less often are responsible for finding their own food and shelter; whereas food and shelter for weekend (or longer) retreats are generally provided for attendees as part of the attendance fee.

Free on-farm events, in which you're not charging a service or per-head entrance fee, are often created to promote the farm or lead customers to direct sales of on-farm products. Consider the possible amount of gross revenue from products sold during the event and if you'll have enough available during that event. Later, with the help of following chapters, you'll factor in further expenses, unforeseen added costs, and subtract the costs of insuring and hosting the production.

Numerous successful agritourism farmers report that benefits "down the road" for agritourism projects or events are priceless and often impossible to calculate upfront. An open house might lead a visitor to tell a dozen others of your CSA farm, and generate new customers next year. An open house of your B&B during a Home and Garden tour could lead to bookings by visitors' relatives six months later. Allowing your sisters' quilting group to set up shop one Saturday afternoon for free (because you owed her a favor) may end up with requests for longer weekend hobbyist retreats.

Farm museums, featuring items like antique farming equipment, work well for agritourism.

tomato, sweet corn, and vintage melon tastings are also very popular.

- *Bird-watchers' gatherings. Agritourism and Nature Tourism in California* reports that bird-watching is more popular than hiking, camping, fishing, hunting and golf, and that bird-watchers are some of the most affluent of all travelers. Invite a member of your local Audubon chapter to discuss local birds in your area along with a workshop on making birdfeeders or birdhouses with your birdhouse gourds; then invite the birdwatchers to sign up for an early morning, bird-watching safari on your farm the next morning.

- *Cultural heritage celebrations* inviting local history buffs to speak, or historical re-enactments, ethnic celebrations and renaissance fairs.

- *Demonstrations and workshops:* cheese making or goat milking or soap making, etc.

- *Fiber arts workshops* including carding, spinning, weaving, rug hooking and knitting. Knitting has recently gained a surge in popularity again among people as young as grade school.

- *Old-fashioned rural and nature crafts:* whittling, knot-tying, making corn-husk dolls or yarn toy animals, quilting, basket-making, wooden toy making.

- *Workshops on caring for animals.*

- *On-farm museum of family, local, or farm antiques.* This does not have to be elaborate. An old shed laced with barn swallows (well, except for nests right above the door entrance) offers a sense of hidden or forgotten ambiance far more than a modern museum exhibit. In France, in the area of Dordogne-Perigord mentioned in the previous chapter, the farmer's grown daughter, now operating the farm's bed and breakfast, took us to her dad's two-room shed "museum" filled with the artifacts, fossils and bones of prehistoric people and animals he'd collected off his farm over the years. It was off-the-map, and fascinating. And history doesn't have to go back that far. Great if you own your family's old butter churn and have collected other old dairy farm equipment over the years, but today's children are amazed to see old telephones—the kind that dial and were attached to cords on the wall!

- *Herbal wisdom workshops and retreats* (See Chapter 13 for HeartSong Farm Healing Herbs' unique take on this)
- *Native plant walks:* Invite a member of a local chapter of a native plant society to lead.
- *Workshops* for making cold frames, bat houses or worm beds.
- *A scarecrow parade, fairy house village, or Jack O'Lantern gallery.*
- *Families build scarecrows at a farm workshop* that are put on display for other farm visitors, and are taken home at the end of the event. Same with pumpkin carving. Fairy houses are miniature outdoor homes built only with natural materials found in the woods and fields of the farm, such as moss for carpet, nutshells for bowls, and bark for tables. After a fairy workshop and house-building event, farm visitors can walk through the woods looking for miniature homes built in nature. In parts of New England, fairy houses are built by adults and kids alike every year.
- *Bonfires or camp fires.* People are drawn to fire and rarely get to experience open outdoor fires anymore. When added along with cornfests or apple pressings or other crop specific celebrations, they can be used to demonstrate roasting apples or corn over coals. Employ all safety cautions, of course, especially if children are around. Urban children especially often have less common sense when it comes to natural phenomena like fire.
- *The familiar hay rides, sleigh rides or corn mazes made unique to your particular farm:* A banjo player on the hay ride, treasures or spooks found amidst the corn maze, a sleigh stopover to one of Santa's elf workshops.
- *Plays or talent shows in the barn.*
- *Children's face-painting based on farm themes:* butterflies, ladybugs, strawberries and flowers. Local art teachers can sometimes find older art students to volunteer or work for hire.
- *A mystery or whodunit party* with the sheds and troughs and barn as locations to find clues and hints.
- *Storytellings* by someone dressed in character (cowboy, Anne of Green Gables, Kokopelli) along with your taste-testings and bonfires.
- *Migratory bird festivals,* e.g. return of the hummingbirds.
- *Mountain bikers' stopover and trails.*
- *Farm memberships* beyond just picking up CSA food shares: Adding a fee option for guests to visit the farm at their will in a self-guided manner to enjoy the scenery, paint, meditate, picnic, photograph or journal, during specific hours when only members are allowed.
- *Somewhat related to the idea above, farm "clubs:"* The idea of being part of a club is appealing to both adults and children. The owners of a community supported agriculture farm, for example, called their CSA a "Food Club," operating it much like a regular CSA. Another example is a naturalists' club. Members gather monthly to observe and record the changing flora and fauna during the seasons on the farm and conduct nature experiments, such as setting up a weather station on the farm. This hobby is healing and meditative for adults, and fascinating enrichment for children. The titles, *Celebrating the Earth* by Norma J. Livo and *Exploring in Backyard Biology* by R. Gary Raham detail the value of being a hobby naturalist. Both titles reveal the reasons and how-tos for keeping naturalists' notebooks and journals, including how these tools develop both hemispheres of the brain and how recording and drawing our observances can advance our intelligence. People learn to see patterns in the world around them, learn the techniques of field observation used by scientists and writers, gain a desire to protect the natural world, learn to keenly observe and remain "grounded" (which some say is the opposite of Attention Deficit Disorder),

Surfing Goat Dairy: Examples from a successful agritourism farm of their pricing and innovative agritourism offerings

Here's a wonderful success example of a dairy with activities unique to the farm, taking advantage of well-known holidays, and using pricing that works according to the farm's unique location.

In the beautiful islands of Hawaii, there is a surprise just beyond the sugarcane fields that line the slopes of one of the island's volcanoes. Lush green pastures can eventually be seen, which host Surfing Goat Dairy, one of Hawaii's only two goat dairies. Here, the goats are free to forage in the tropical sun. The owners use no hormones, antibiotics, pesticides, herbicides, preservatives nor colors in their cheese production. Numerous artisan goat cheeses, many which have won national awards, are produced on the dairy, and agritourism plays an important role in the farm's operation. In fact, they've turned their farm into a top agritourism destination. During the events, their Cheese and Specialty Shop sells more than 25 different cheeses, which they always have available for tasting and purchase, along with light snacks and beverages.

Their agritourism is comprised of several special events along with ongoing regular activities.

Surfing Goat Dairy, www.surfinggoatdairy.com

"Aloha from Maui," says farm owner Thomas Kafsack. "For our bigger events (more than 1,000 visitors) we have the problem that our parking space is limited. That's why we price them by the car load, which helps to reduce the number of cars. And it works – we got only three phone calls (the last time) about single persons who didn't know anybody. The prices for all of our regular tours are calculated by the length of the tour and if there are cheese tastings, food or beverages included."

Here's a sample of Surfing Goat Dairy's regular ongoing activities and prices (at this writing):

♦ "Casual Tours," offered daily, no reservations needed ($15 up to 3 people – $5 each additional person).

♦ "Evening Chores & Milking Tours," offered Tuesday, Thursday and Saturdays at 3:15 p.m. ($10/pp).

♦ "Grand Tours," First Saturday of the month for two hours. Visitors can feed and milk a goat, see the cheese making process, and sample most of the cheeses produced at the dairy ($25/pp).

♦ Special events have included a Farm-to-Table fun fair where well-known chefs are invited to give cooking demonstrations. For this event, the owners charged $20 per carload.

♦ Special Valentine's dinners, and Father's and Mother's Day brunches priced differently each year.

gain stronger sensory skills and quiet their minds, among many other benefits.

- *Event rental location.* Some farms are open to just about any event that's legal, and basically stay out of the picture completely. Others lean towards marketing to attract specific types of events, still staying away from the event themselves, or sometimes getting a little involved.
- *Culinary retreats* with cooking classes featuring the farm's produce. Some farms arrange their own; others partner with a chef (see Chapter 15)
- *Recreation equipment rental* when the farm is near, or encompasses, special nature recreational opportunities: regional birding books, binoculars, canoes, tents.
- *Special attraction gardens* (See also: Chapter 12 for more detail on how farms use gardens to attract customers).
- *Tea gardens.*

- *Children's theme gardens:* alphabet garden (again, see Chapter 12), Little House on the Prairie garden, pizza garden;
- *Historical gardens:* a Shaker medicinal garden, a Biblical garden, a Native American Three Sisters garden or colonial garden.
- *Ethnic gardens:* Italian, French, German cooking staples grown in the country's current garden style.
- *Cut-flower garden* featuring tulips, dahlias, sunflowers. These are especially attractive near roadside stands.
- *Natural dye garden.*
- *A roadside stand with changing seasonal themes* or with a specific theme such as Old West or local heritage.
- *Photo booth* with costumes or even a very patient pony to borrow for the photo.
- *Unique farm animal viewing.* A U-gather filbert farmer allows his peacocks to wander amidst the grove much of the year. Before and during the

Visitors may enjoy replicas of exotic gardens, like this one growing in a castle courtyard in France.

CHAPTER TWO

Agritourism Customer Service: Beyond the "Smiling Face"

While it is beyond the scope of this book to summarize the enormous amount of written material already available on the subject of customer service and obtaining customer feedback in order to improve your business, business literature of the past several decades has laid huge emphasis on the value of repeat customers. One study, for example, states that it is five times as expensive to find new customers through advertising or promotion as to retain old ones!

Customer service also is huge in agritourism: You simply can't do it if you don't naturally like people. Agritourism is, in fact, inherently such a people business that a smiling, put-the-customer-first facade will not work with agritourism as it might for a few hours during a farmers' market. If you can't stand people, you most likely won't be in the agritourism business in the first place – it isn't for everyone.

To choose your activity, you need to know your own personality as far as working with people, as well as how agritourism might fit in with your overall farm vision.

Do you like to work quietly while others respectfully observe? Maybe tours narrated by another family member would work for you. Do you like being the charismatic shopkeeper occasionally? If so, perhaps an on-farm shop will be your choice. Have you always wanted to host? Teach kids? Cook gourmet? If "yes," this might help you decide that agritourism is for you.

Finally, it should be added that customer service and hospitality, in agritourism, often assume a different personality than they might with a large roadside stand or farmers' market. Nikki Rose, of Crete's Culinary Sanctuaries described in Chapter 1 and the Epilogue, reminds us that farmers are not our servants who owe us entertainment or royal treatment. She and others agree that one of their closest descriptions beyond farmer, when it comes to agritourism, is "educator." They must learn the ropes of working with people, but in return, they have the right to expect respect from their "students" and dismiss those that are disruptive. The author was on a recent farm tour when the farmer got a sudden call that

the local co-op could use 16 more flats of berries. They apologized for making us wait an extra 15 minutes while they loaded the berries up and one of them drove the berries in, while the other took us on the tour. It was more authentic and exciting to see what really happens on a real farm, and to let the farmer continue to be a farmer even though we were there. It's a delicate balance, allowing the farmer to be a farmer, but be friendly, and keeping him or her from becoming "entertainment first, farm second."

There really are times when the farmer needs to just work the land or care for the harvest while others observe, even answer phones or take care of the goats first, and that's what the tourists come to see. And, it would be exhausting to have people coming to your own farm home on a regular basis if you don't like people naturally or felt you had to stop your chores and treat them like royalty.

Resource: For customer service tips specific to agricultural operations, see "Customer Service" on New World Publishing's website at www.nwpub.net (go to "Free Downloads").

U-gathering season, he puts the birds in a large nearby viewing pen. Kids, especially, enjoy finding the beautiful peacock feathers while gathering nuts and then viewing the birds up close. As with all activities, check local regulations. There may be requirements or suggestions on how long live animals must have been removed from areas of food harvest.

Some more specific events include:

- Country weddings.
- *Family reunions* with old-fashioned fun (crochet, horse-shoes).
- *Nature oriented youth camps* (church groups, scouts, summer nature camps or parks and recreation programs).
- *Singles' events:* A CSA farm near Seattle caters to singles, offering share sizes and options that better match a single's lifestyle. They also offer larger shares for couples and families. Singles find themselves meeting other eligible singles of like mind through this CSA, and the rest is history.

When farmers offer their location as a rental site and find themselves attracting and enjoying certain events more than others, such as "Country Weddings," some add further agritourism opportunities related to that event.

Here are some wedding examples:

- *A U-pick, bridal cut-flower garden.*
- *Bridal flower arranging demonstrations.*
- *Dried lavender bundles* to open and toss over the newly married bride and groom (replaces rice or birdseed).
- *Bridesmaids' pre-wedding tea* or on-farm bridesmaid herbal manicure and pedicure retreats. As with all projects, check with local authorities to see if special rules apply or licenses are required. Some states regulate professional facials, manicures and pedicures. If so, see if you can legally replace them with "herbal foot and hand soaks" or make a trade with a licensed professional.
- *Pre-wedding bridal party dinners or picnics.*
- *Bride and groom shower location rental.*
- *Carriage rides down a flowered path for bride and/or groom's entrance.*
- *After wedding country entertainment* for wedding guests: Pony rides, wagon rides, cooking demonstrations, straw bale maze, petting zoo, garden tours.

Near-explosions in word-of-mouth promotion can occur when a specific theme is adhered to. One bride may have a church wedding already planned, but will come out to pick her wedding flowers at the farm, witness all that's available on your farm, and spread the word to others in the wedding planning stage.

Ideas that work when seeking overnight guests:

- Design and print brochures with a list of other local attractions, especially if you are planning to attract overnight guests.
- Set out a notebook for collecting, over time, a list of already-established, local attractions that may attract guests to your farm and make their stay with you even richer. Once you make up your mind to find these attractions, you may discover that what you once thought was a very mundane location is wealthier in possibility than you previously thought. Once you start looking for these attractions, they seem to just "show up." The grocery clerk, the junk mail, or a comment or remark overheard while waiting to get your tires rotated, are all potential sources for ideas.

Fill your notebook with the following:

If you're close enough to a town:

- Dates and locations of local parades, light shows, art walks, home and garden shows, plays, storytellings, outdoor movies in the park, local music events, and walking tours of historic districts.

(Read your local newspaper for event listings and gather brochures at your local Chamber of Commerce).

- Attractions put on by local businesses, such as tours of processing plants or behind the scenes of the movie theater.
- Antique shops, used bookstores, French bakeries, massage therapists, natural product day spas–anything that's unique and not franchised.

For those both near town and far out into the country, look into:

- Native birds, both resident and migratory.
- Natural attractions, such as nearby river beaches, kayaking, trails.

- Description of the local history: Was your canyon once a huge prehistoric river? Is there an interesting Native American or First Nation's story to your area and are there nearby gift shops or programs visitors may want to take part in? In our county, for example, our local Coast Salish Natives hold Earth Day celebrations, canoe races and dancing demonstrations where the general public is invited.
- Who were the first settlers?
- What natural formations might interest amateur geologists?

Once your list is complete, include it in your promotional website and brochures, and keep them updated. ❧

Part II: Starting and Growing Gracefully

The Business Plan 3

Starting accidentally on purpose

Several successful farmers interviewed throughout this book started their agritourism enterprise "accidentally." One farming couple, for example, was asked by their church to set up a haunted barn for a church fundraiser, and their agritourism business grew from there. On MaryJane's Farm in Idaho, daughter Megan was getting married on the farm, and the family set up wall tents in the old orchard to entice wedding guests to stay on longer. These grew into a rustic bed and breakfast agritourism project with guests enjoying the tents along with outdoor kitchens. On Spoutwood Farm in Pennsylvania, a CSA farm, the owners Rob and Lucy Wood held a springtime Celtic-themed party for about 100 friends. It grew from a one-day party to a two-day festival hosting close to 10,000 people, complete with Celtic musicians, artist and food vendors, and an entrance fee of ten dollars per adult and five dollars per child age three and up.

But, ironically, these businesses were in some ways accidentally following the plan that successful, "planned out" enterprises follow on purpose. Farming, including agritourism, calls for direct experience, and few farmers can leave their operations to apprentice for several months or a year. So, four stages similar to those that follow have been suggest-ed by various farm advisors, including *Stockman Grass Farmer* editor, Allan Nation:

- Acquire the nuts and bolts knowledge needed to produce and market the farm's new product or service on your own time (free, local ag-extension workshops, books, online networking with others in the business, etc.);
- Provide or produce it for yourself and your family;
- Provide or produce it for your friends who have tried it and ask for more;
- Turn it into a business.

As a fictitious example, Meadows Farm has an old, unused cabin on the property. The owners study books and attend workshops on old home restoration, talk to a friend with construction experience, and look into operating B&Bs in their area. Once cleaned and spiffed up with new porch railing that meets local codes, grown visiting children and an aunt and uncle are allowed to spend the night there. After addressing complaints about spiders and fixing a slippery stair step, visiting friends spend the night, complete with breakfast in a basket delivered on the porch the next morning. Several ask to visit again. The following summer, the owners offer it as a paid B&B farmstay. The first customers come from word-of-mouth via the original friends who enjoyed their stay, and the press release sent to the local newspaper which generates several more guests throughout August and September.

Another real-life example is a gourmet food garden that first fed the family, then produced gifts of fresh produce and preserves for friends and extended family, and now offers farmed products for sale from the farm's own roadside stand.

The practice stage is also the time to start taking quality color photos for future promotion once it's the real thing: your flower fields in bloom with children picking flowers, groups on hay rides, people entering your restored cabin, etc. Find a friend with experience or a natural knack for artistic photography, if you don't possess it yourself.

If you're able to start slowly this way and gain experience before putting out a lot of start-up money and risk up front, you may find you've built a strong foundation that will leap frog you into a successful future, whereas jumping in as a business immediately may leave you feeling a little shaky.

Whether you must start as a business immediately, or are able to first practice-operate your enterprise, a customized career roadmap will raise the odds of more and faster success and help you continue forward. Note the real-life example below of the North Dakota Farm owned by the Engers. They refused to jump in too quickly, and eventually created a solid destination farm visited by people from miles and miles away.

Drawing a roadmap for success

Some businesses grow and progress year after year. Others barely succeed and always seem to be chasing emergencies. Finally, of course, there are those that outright fail. This chapter is about using a process that many of those from the first group use. They all make the same type of roadmap that they follow to succeed, even though each roadmap is different for each business. A roadmap is indispensable when starting a journey towards an unfamiliar destination. It allows you to use a well-established method for tinkering with your farm's agribusiness element on

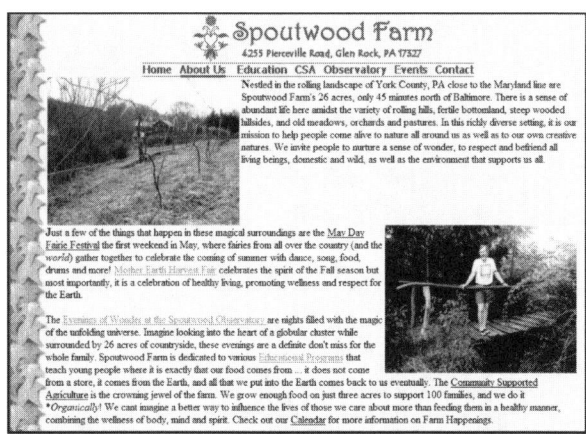

On Spoutwood Farm in Pennsylvania, a CSA farm, the owners Rob and Lucy Wood hold a springtime Celtic-themed festival hosting close to 10,000 people.
Spoutwood Farm, www.spoutwood.com

paper before you even begin, trying on different approaches before taking risks, and then going forth with a roadmap now aimed at your target that no one but you has access to. It then later allows the roadmap to become an operating tool, continuing year after year to help you stay far ahead when it comes to giving paying customers what they're looking for.

Making the roadmap

Whether you've been in the farming business for years and are adding or enhancing agritourism now, or are planning a brand new farm with an agritourism element, the time-honored way to either improve, or start off ahead and raise the odds of success, is to write a custom business plan which creates your farm's unique, how-to book, and a roadmap to where you plan to be over the next two to five years. It gives you a bird's eye view of your plan so you can try adding one more harvest weekend to see how much better the numbers add up, or try out another customer outlet, such as a chef's school, to attract farm visitors with fliers of your on-farm culinary events. It gets all your thoughts and energies focused, targeted and organized, and makes you think of what you hadn't thought of before when you simply

CHAPTER THREE

kept the idea in your head or jotted down a few notes. It's as though it accesses an area of your mind that successful business owners naturally utilize, and keeps that area of your mind open and operating.

> "The aim of developing a business plan is to help you strive towards the opportunities of tomorrow. It is a process used by those who aim for continued improvement and for those who wish to make more of better things, with less resource, less cost and less human effort. Business plans help maintain the difference between mediocrity and superiority. That is why we are here, isn't it?"
>
> –Trish Murphy of the New South Wales, Australia, Department of Agriculture

Proponents of farm business plans remind us that it's possible to be very, very busy, and to work very, very hard, and still not be efficient. While business plans may not have been typical farming tools of the past, including the farm crisis of the 1980s, they are becoming more widespread among farmers of all types.

"When producers go through the exercise of developing their business plans," says Amanda Snyder, Sustainable Ag Program Associate at the University of Idaho, "they can identify potential challenges, constraints, and pitfalls. Developing a plan may help reduce the risk of falling too far in debt to make up for costs that were not expected. Having a draft business plan may allow producers to view what options they have not considered before, and provide a long-term outline for their growing enterprise. Issues that need to be planned for when developing an agricultural business might include planning to finance or rent equipment, how and when to purchase other supplies, how to manage hired labor, how to budget for health care and other liability concerns, as well as planning for additional input costs." Amanda is involved in teaching the Agricultural Entrepreneur course described more below and so is Todd Murray of the Washington State University Extension near Seattle. He reminds us that a

business plan can greatly improve a farm business, even if the farm has been in business for years. "It's never too late to make a plan," he said. "In our Agricultural Entrepreneur classes we have well established farmers coming to finally write up a working business plan. Even after years of running the businesses, growers can use the plan to see where they can improve and refocus their efforts. A good business plan will help identify and adjust for opportunities."

For some who still resist, you may worry that your initial enthusiasm for a new or expanded business will die once you get down to specifics and calculations in a business plan even before beginning: "Why bother crossing that bridge until we get there? It's smarter to wait until we're buried in it; then it will be too late to turn back, and we won't run the risk of just giving up before we've even started. Besides, I don't know all the answers up front, yet." Well, the time-honored business plan has worked for years, even though nobody has ever really known all the answers up front first. Its very questions have a way of opening new doors, of business discovery and self-discovery one had never thought of before, of resources, money-making ideas and cost-cutting possibilities never visible before. It's much easier to make changes on paper first when we find that bridge we were waiting to cross led to a previously unknown cliff's edge. Before making such a plan, you were inspired. But once the plan is completed, you're again inspired along with being confident and laser-focused.

While some business plans might be written solely to obtain financing from a bank, business plans have value far beyond impressing bankers. It's easy to wish there was already a written roadmap for your particular venture, but there is no cookie cutter plan customized for you unless you want to be an exact replica of the next business. If your plans fall into some of the more basic agritourism projects: picnic tables, wagon rides and petting zoos, your ag extension may have some pre-made scripts to follow for these projects. They can be very helpful. But you're adding these components into an overall plan that is yours alone. Your own business plan is the creation of your own cookie cutter, one that makes only you succeed.

Getting started

First, know what you'll use your business plan for. Your own private roadmap for you and your farming partners? Convincing an institution or individual to offer funds? Not sure yet, hoping the business plan will reveal its best uses? You may want to write a basic plan and then customize several variations for more than one audience if new audiences become clear during the planning. In this chapter we'll discuss five ways to create your business plan.

The first way is to simply do it on your own, for free and at your own pace. Obtain a basic business plan template online—which is basically a series of categorized questions for you to answer—that's written for all types of businesses. These are free from several sources (see "Resources" sidebar in this chapter). If such a thing exists, a template written for a business somewhat similar to yours might be even better. While some business plan templates are written for specific businesses, including "farmers," a five-acre destination lavender farm B&B may have more in common with a country inn's business plan template and less in common with that of a 500-acre cattle ranch.

A basic business plan template may include the following sections, sometimes using different names for the same section, such as "Vision Statement" instead of "Mission Statement." There are questions within each section to answer:

Executive Summary: This is a one or two-page summary of your overall plan. Although it is eventually the first section to be seen, it is often written last. That's because it's actually a combination of two other sections not yet written, which are the Business Concept section and conclusions from the Financial Strategy section.

Mission Statement: In 50 words or less, your mission statement will focus the core purpose of your business in a manner that reflects your higher values and goals. As an example: Island Farm will help restore our area's sustainable local farming heritage, preserve rural character, feed our local community, and contribute a new tourist attraction. By networking with local businesses and schools, growing eco-tourist populations will experience unique island flavor, while locals learn sustainable farming and gardening through festivals and workshops.

Business Concept: A one page persuasive description of the who, what, where, why, when and how of your farm and agritourism enterprise, including your precise products and services, target customers, how it fits in with your current farm if such exists, and how it complements and fairly competes with other local businesses. Since it is also the first part of your Executive Summary above, and that should also only be one page, make it a "short page."

Products and Services: These are sometimes described entirely within the Business Concept Section above (i.e., organic antique apples and an annual apple festival). But they may need their own section if your products and services are wide and diversified.

Development Goals and Objectives: Includes long-term goals, that which you want accomplished in the next three to five years, and shorter-term objectives which are the methods by which you will reach those goals.

Background Information: For this section you may be answering why your products and services will do well, and include statistics on growing eco-tourism trends worldwide and in your location. It may go deeper into local complimentary businesses which will show how your business already has much leverage to succeed, and even deeper into why you will succeed in spite of any competition.

Management Profile: Some say that investors will look at this section even before they look at the quality of the product, with the idea that even mediocre hamburgers and fries can succeed and spread across the world if there's a good management plan backing up those products. How will you assure your business runs smoothly? What is your business entity? Who are you insured with? How will you evaluate further growth? Who is on your management team: Do you employ a good CPA, or other professional advisors, for example? What are the regulations in your area and how are you handling them?

Marketing Plan: What are all the ways in which you'll promote your agritourism business, and how will you evaluate these strategies?

Financial Plan: This section will state where you are now, financial forecasts, what your new needs are, and which you can play around with and manipulate until you feel satisfied.

Appendix: This section holds supporting documents, such as support statements from your current customers, financial statements, and the like.

Even if sections have different names in different business templates, most of the above issues are eventually covered, even if one section ends up splitting in two. The SCORE business template, for example, includes a General Company Description in which you include your mission statement.

Again, business plans meant especially for farmers can make the process more familiar and more customized, and may add, remove, or reorganize questions. Amanda Snyder, when asked what other areas such as Executive Summary, Mission or Vision Statement, and Management Profile, are covered in the course she helps teach farmers on creating their own business plans, replied: "Some of the other topics we cover in our courses include Diversifying an Agricultural Business, Goal Setting, Product and Industry, Market Analysis, Marketing Plan, Raising Capital, Management and Operations, Record Keeping, Planning for Profit, Financial Statements

Analysis, Budgeting, and Cash Flow Spreadsheets (providing examples of these)."

Next, see if you can find someone else's completed business plan to look over. Seeing an actual example of a plan will usually make the writing of your own make more sense. While most business plans are kept confidential, see the Resources sidebar in this chapter for possible locations for completed business plans. Also, simply search online for "agricultural business plan," or even for something similar to what you have planned, such as, "bed and breakfast business plan." At this writing, there are several plans displayed online through searches, but these locations change month to month, old ones disappearing while new ones appear.

Once you've found an appropriate template and have seen others' plans, go over the template in entirety first, and note how you'll customize it just for you. For example, if you don't have employees, skip any parts that pertain to them. Then, proceed to answer, in some cases around 150 questions minus those you've deleted, one at a time.

When finished, smooth out your notes into a readable and convincing narrative. If you plan to offer this to a financial institution or other entity that will evaluate whether to invest in your plan, it's essential that your grammar and spelling are exemplary, as you will be judged by these along with your great business idea.

Show the plan(s) to a farm and business-savvy advisor with experience in successful business plans, such as a SCORE counselor (service is free, see resources below). Make any adjustments they suggest that you agree with.

A second way to create a business plan is to purchase printed books that contain worksheets for your purposes. According to ATTRA, "Two of the very best of these publications are *Farming Alternatives: A Guide to Evaluating the Feasibility of New Farm Based Enterprises,* a workbook from Cornell University, and *A Primer for Selecting New Enterpris-*

es for Your Farm, a Kentucky Extension Service publication. These guides discuss alternative enterprises and introduce a step-by-step process to assess the objectives, resources, markets, production demands, and profitability of new enterprises. Both include a lot of useful worksheets to help with these assessments." Another good workbook is put out by the Small Farm Success Project. Yet another is Building a Sustainable Business: a Guide to Developing a Business Plan for Farms and Rural Businesses which may still be available both as a low-priced print book or free as an online PDF file. See Resources in this chapter and the back of the book for ordering. Again, once complete, show it to a third-party advisor.

A third way to create a business plan is to purchase software that walks you through the business planning journey, checks your financial calculations, and allows you to customize their program to your own business. The Resources sidebar lists one possible resource for such software to get you started shopping. In this case, as with all of your business endeavors, remain in the navigator's seat when choosing products and services rather than becoming the obedient subject of just one business guru (or piece of software). Some software is outstanding, others are little more than a template taken from a free online resource and charging you for it. When a software promotion says it will lead you to further sources of discounted material for businesses, realize that while these may be excellent resources, there may be even better sources out there, and new ones are forming all the time. The software is most likely linking up with discount business product and service companies to lead its readers there for an affiliate fee they're paid each time those who own their software make a purchase.

A fourth way to produce the basic plan with more personalized help from the beginning is to involve a SCORE counselor or other advisor from the start, as did the dairy farmer described in Chapter 16 who not only saved his farm, but turned it into a flourishing business with the help of a SCORE counselor who guided him from start to finish.

A fifth method is to partake in a nearby workshop for creating your own custom business plan. These may call for a less flexible time commitment than you're able to accommodate, such as attending the workshop according to a class schedule, but there are bonuses of networking with others who are involved in farm and home business projects if you're able to attend one of these. Your local community college may offer free or low cost courses for small businesses that walk you step by step through the business plan creation. Check with any courses you see listed if, when you leave, you'll have a completed plan already looked over by their mentors.

You may want to find a business plan workshop aimed at farmers. These workshops allow farmers to gather with one or more mentors who may follow the basics of a business plan, but ask even further, or very different questions which will help you customize your plan even better. They're often put on for free or low cost by universities, the Small Business Administration (SBA), or your local ag-extension. As one example, Washington State University partnered with the University of Idaho and the Small Business Development Centers in both states to offer a course entitled Agricultural Entrepreneurship, the one mentioned above assisted by Amanda Snyder and Todd Murray, where the student's final product is their own business plan ready to take to the bank or to use as their own roadmap. The course also leaves students with a networking list of professional and peer contacts for future support and advice. Its price has been around $200 for 12 weeks, with scholarships and college credits available. Sean O'Neill, instructor of the Agricultural Entrepreneurship course, said this course is available annually. "We offer this course at the University of Idaho (UI) each spring semester. It is sponsored through UI, Washington State University (WSU), Rural Roots, and the Small Business Development Centers (SBDCs), among others. We have academic students from UI and WSU as well as community members that take the class without formal academic registration. I took the course as a student in 2002, so it's been offered for some time now."

The Minnesota Institute for Sustainable Agriculture's Whole Farm Planning program is another example of a program set up to help you produce a business plan. It has turned the business plan into a four-step process, as they describe, "which can be used by the farm family to balance the quality of life they desire with the farm's resources, the need for production and profitability, and long-term stewardship. How is Whole Farm Planning different from the planning you already do on your farm? Most farmers do some kind of planning on their farms nearly every day. Whole Farm Planning is distinct from other farm planning approaches because it ties all the planning you do together for the whole farm and bases it on the long-term vision your family has for itself and the farm in the future. It is farmer controlled, voluntary, and flexible. The plans are owned by the farmer and the information contained in the plans is confidential."

To find out if anything similar is available near you, look for one of the hundreds of Small Business Development Centers (SBDCs) close to you, also sponsored by the Small Business Administration (SBA) in partnership with educational institutions, state and local governments and the private community. They provide various resources and on-site counseling for start-up and expanding businesses. To locate a close-by SBDC, contact the SBA Small Business Answer Desk at 800-827-5722 or find them online.

Continued evaluation and feedback: Turning the business plan into an operating plan

Once your business plan is complete, and has been used for the initial purpose you've chosen such as bank financing, your own private start-up plan, or getting a grant, your business plan now turns into your operating plan. This basically means it be-comes a tool you use as you continue to get feedback, evaluate, update, tweak, and improve the original plan as an ever-evolving guide and roadmap towards continued success. In the Agricultural Entrepreneurship course, students are taught that the business plan has a continual purpose. "The use of the business plan is ongoing," says Sean O'Neill. "Periodically, particularly in the early stages of the business, actual numbers replace numbers that were originally used in budgets and financials to facilitate future planning. The plan is also important as a roadmap for the growth of the business and as a communication tool for employees. As the external and internal environments evolve over time (new opportunities and challenges) it's important to go through the process of seeing how they'll affect the business in the big picture. The format of the plan ensures you cover all your bases."

To progress and grow, you will need to make ongoing efforts to keep and generate agribusiness customers, so find ways to allow customers to offer feedback to you, privately and anonymously. Actively seek customer feedback and consider it as valuable indications for further improvement, rather than criticism.

It is very valuable for farms to read customer feedback such as, "There's already too many corn mazes around here. How about a corn labyrinth instead?" Or, "I'm using your farm to expose my kids to something besides constant competition. Could there be a pumpkin carving demonstration and then a pumpkin art gallery showcasing all the kids' pumpkins, instead of the usual pumpkin carving contest with only one winner?"

Feedback from customers, as well as feedback you observe yourself on where humanity is heading and how that fits in with your farm, will be one of the most valuable methods of guiding your future business destination. Again, farmers inherently understand natural selection. It's the same with agritourism projects. Which ones work better than others and can be saved to produce even more customers the following season? Which ones mutate into something unuseful? Which ones mutate into something better than originally planned? How can you expand on the best mutations?

Naomi Karten of Karten Associates, delivers seminars and presentations internationally to improve customer satisfaction, manage change, strengthen teamwork, and improve communication and consulting skills (See Resources). Naomi states that it is highly valuable to give customers a convenient way to offer feedback with plenty of space to write their own thoughts, and this is often overlooked by the businesses she's come in contact with. As seen in Chapter 16, Sweet Grass Dairy's successful cheesemaking classes were a result of feedback requests written by customers during tours of their dairy farm.

Even customers' slightest offhand comments may give you a break-through idea that can have them eagerly seeking your farm's experiences. Methods of obtaining feedback may include blank sheets available at your first open house or handed out to your CSA customers, survey sheets handed out at homeschool co-ops, food co-ops, gardening clubs, or a list of people you've compiled on your own if you haven't started an agritourism project yet, or even an invitation for suggestions via an article written up about you in your local newspaper (see Chapter 9), inviting customers to go to your website and fill out an online suggestion form.

Naomi's suggestions for ongoing successful customer feedback include the following: Make customer feedback a process, not an event. In other words, think ahead on why you're getting feedback in the first place: how it will help the farm progress. Develop a system for gathering and studying feedback. And then, once received, do something proactive about it rather than simply using it as fodder for complaining about city slickers at the dinner table. Naomi said she too often sees customer feedback treated as something that "just happens" or as an "event" such as e-mails received from customers, without using it as a valuable tool for business growth.

"The activity is often treated as an end in itself rather than as a means to understand and respond to customer needs," she states. "For example, one company I visited had conducted a customer satisfaction survey which revealed that customers were dissatisfied with certain aspects of the group's performance. When I asked the manager what his customers actually meant by this grievance, he said he didn't know and confessed that no follow-up had been done with customers to learn more about their complaints. Worse, no steps had been taken to address the concerns customers had raised."

Starting the process of gathering feedback

The starting point in developing a feedback gathering process is to first address the following questions:

What are your reasons for taking feedback? Is it to assess customer satisfaction to see how well they're enjoying your farmstay, or if they truly like picking blueberries on the farm enough to come back again and to tell others? Naomi states that assessing customer satisfaction is the most common objective, but it is not the only possibility.

Other reasons are to keep tabs on what customers describe as important to them about their time spent with you. Although you can tell they enjoyed their farm tour, you were certain it was the llamas that drew them there, until a feedback sheet revealed that for the children, it was the horses. Over time, patterns may emerge that you hadn't thought of before. Perhaps you started your agritourism project as a rustic B&B with no TV or Internet hook-up for customers, with full intentions of adding those options down the road as revenue allowed. But if customers continually state they enjoyed the freedom they felt in having no phone or TV, you can take note of that down the road when considering upgrades. Had you simply allowed your customers to rate their satisfaction level between one and ten, the exact issues important to them may not have been revealed. Maybe you'll find that customers look forward to a brand new type of monster to scare them in the haunted barn every year… or just the opposite, that the pony you had dressed up like a reindeer for your Christmas tree farm is something many families look forward to year after year, and hope will never change.

You may want to know what's changing in your customers' environment that could affect your ability to serve them. Are their children's schools becoming increasingly commercialized? If so, this can set in trends among parents to seek outside enrichment for their kids, such as programs involving nature. Are their teens beginning to respond differently to farms and rural traditions than past teens? (Knitting, for example, has recently been shown to have soared in interest among girls and teens!)

You may also want to periodically reassess whether your customers really understand the nature and scope of your services. Farmers who succeed in agritourism continually state that the people coming to their farms are appreciative, curious about, and supportive of farming, and the farmers don't want any other type of customer. If your customers continually explain that their children were bored during their visit with you, you may want to redirect your promotion towards a new audience with a higher chance of involved children whose parents instill a love and fascination for nature and the non-commercial world, such as scout groups or homeschooling groups.

When is it beneficial for agritourism farms to gather feedback? Naomi's suggestions include the following:

At the beginning of something new, either with new customers (such as a CSA opening up once a month to the general public), or a new agritourism project, such as adding heirloom garden vegetable starts to your already established spring open house offerings. This might mean setting up a feedback corner in the greenhouse where the starts are sold, with paper, pens, and a box for dropping in feedback.

At pre-selected "checkpoints" during a lengthy project, such as automatically handing out surveys during the middle and end of the harvest seasons for U-pick or CSA customers.

At the first sign of customer dissatisfaction, such as when the number of farm visitors seems to be dwindling, and you decide to leave survey and feedback forms in the rooms of your farm's B&B to discover what the customers are liking and disliking.

Gathering the feedback

There are several ways to gather feedback, both formally and informally. "Focus groups" are sometimes quite formal when used within larger businesses and may not seem appropriate for you. But focus groups and "learning clusters" are used successfully by larger entities you may be involved with such as your local Chamber of Commerce or a regional advocacy group promoting ecological farming. These focus groups involve a cross section of volunteers who agree to focus on a specific topic and stay on target while they brainstorm and organize solutions and new possibilities. They were used successfully by the Association for Enterprise Opportunity (AEO) when studying how micro enterprises, including agritourism, could help revive rural communities. Within your own farm business, you may find it useful to initiate less formal focus groups. Ask several members of your CSA to gather one afternoon to focus on how you can better serve their children, or ask another group to explore gourmet food trends the farm could possibly address.

Holding periodical meetings or even gripe sessions may be more your style. Have monthly pot lucks either with farm customers themselves, or farm family and farm workers who intermingle with the customers, to hear what they've learned from their interactions. Even casual chats with customers are very important segments of the feedback system, and you may want to keep a notebook just to jot down comments you heard as the result of direct conversation, or those you overheard. Naomi states

Resources for business plan templates, sample business plans, business plan workshops, and business advisors and mentors

SCORE, administered by U.S. Small Business Administration, calls itself "Counselors to America's Small Business" and has business plan templates and sample business plans.

http://www.score.org

National Sustainable Agriculture Information Service (ATTRA)

http://attra.ncat.org

Sustainable Agriculture Research and Education (SARE)

http://www.sare.org

Your own local community college or university agricultural extension

Agriculture Business Plan, and Business Plan Template (from The Green River Community College class on Developing A Business Plan).

http://www.greenriver.edu/busines scenter/resources.htm.

Your **State Departments of Agriculture** may also have farm related business plan templates and guidance online for building the farm business.

ww.MoreBusiness.com, website for entrepreneurs, has free articles for creating business plans, sample completed business plans, software you can purchase for business planning, and more.

The Small Farm Success Project, more information in resources in the back of the book.

www.smallfarmsuccess.com

BizPlanIt has a free online business plan template, as well as software for sale.

www.bizplanit.com/vplan.html

Building a Sustainable Business: a Guide to Developing a Business Plan for Farms and Rural Businesses
Publication from SARE that helps alternative and sustainable agriculture entrepreneurs to develop profitable enterprises. Sample worksheets illustrate how real farm families set goals, researched processing alternatives, determined potential markets, and evaluated financing options. Blank worksheets help producers to develop detailed, lender-ready business plans

and map out strategies to take advantage of new opportunities. Print copies are available for $14 (plus $3.95 shipping and handling charge), or the publication is available free of charge at:

www.sare.org/publications/busines s/business.pdf

To order print copies, contact: Sustainable Agriculture Publications, 210 Hills Building, University of Vermont, Burlington, VT 05405-0082. PH: 802-656-0484. FAX: 802-656-9091.

sanpubs@uvm.edu

Naomi Karten, Karten Associates
Web site includes free informative newsletters on succeeding at business.. Naomi's print books and e-books include "Managing Expectations: Working with People Who Want More, Better, Faster, Sooner, NOW!", "Changing How You Communicate During Change" and "How to Survive, Excel and Advance as an Introvert."

www.nkarten.com

that a well-thought-out feedback process includes a combination of methods, each used where it will be most effective.

Make feedback convenient for customers to offer. As stated by Eric Gibson, author of *Sell What you Sow!,* "Make it easy for your customers to communicate with you. Provide postcards or evaluation pads (with pens or pencils available), similar to those offered on the back of restaurant checks, for constant evaluation of your services. Place a large sugges-

tion box near the checkout stand (if you have one) for customers to place them in. Use phrases like: How can we serve you better? Find out what customers like about you, don t like about you, and what they wish you would offer them. The customers' wish list is your key to new sales opportunities!"

One of the worst mistakes Naomi saw of a business' survey technique was to send customer feedback sheets home with customers, and then expect them to affix a postage stamp and mail it back.

While this method certainly allows the customer time and privacy, few people remember to follow through after they get back home, nor appreciate the expectation of paying for it themselves, regardless of how little a postage stamp costs. Provide a postage paid feedback form as an option if you want, but also provide a feedback corner on the farm itself when possible: a private, comfortable writing area and a slotted feedback box where customers are welcome to say what they want without having to sign their names. Put up a visible sign for customers to know this feedback area is available. Typical signs read, "How are we dong?" or "We value your feedback," or "Got thoughts?"

Naomi advises giving plenty of room for the customer to express themselves rather than just a checklist. A simple list including: "Did you like petting the lambs…yes, no," won't convey a lot. She reports that she sees far too many surveys by businesses that don't allow room for the customer to explain themselves in ways the business doesn't know how to ask, nor to choose what they, themselves, find valuable. If you allow customers to pick their own flowers from one of your cutting gardens, they need open-ended survey questions such as, "What was the most enjoyable reason for your visit?" in order to state, "We loved taking photos of our kids for our family Christmas cards with your red barn and mountain in the background, and flowers in the foreground." Had this customer simply checked "yes," or "no," to a survey question of, "Will you come back next year?" you may not have thought of setting up a photo booth in the flower garden and sent in a press release about it to your local newspaper.

Ultimately, however, your continued success will no doubt be a combination of original suggestions from your customers and your own inspirations that no customer would ever think to suggest. If you're purely customer-driven only, you risk being led right over a cliff by customers that represent the status quo, or being too slow to guide your visionary enterprise forward. Potential future customers sometimes have no idea how much they'll enjoy spending time in the sun, picking their own blueberries until they try it, and this pleasure won't show up on a survey asking local citizens who've never picked their own berries before if they want you to set up a U-pick berry patch. But once you find a way to entice them to your U-pick patch with a one-time open-house blueberry festival, complete with a bluegrass band and blueberry ice cream for the kids, they'll be back each summer to harvest, glad you opened that door for them.

Grow slowly and steadily: How one farm did it

—Adapted from USDA Natural Resources Conservation Service, "Alternative Farm Enterprises— Agritourism Success Stories," "Family Education and Entertainment on the Farm," May 2002.

Steve and Dorothy Enger, owners of Enger Farm in Hatton, North Dakota, added a family education and entertainment agritourism project to their 1450 acres of wheat, barley, pinto beans, soybeans, carrots and pumpkins. Below is an adaptation of their own description from a United States Department of Agriculture series of success stories in alternative agriculture. Notice how they carefully added a few agritourism elements at a time to their North Dakota farm to make sure they didn't grow too quickly. It's important to note that Dorothy had at one time been an elementary school teacher, and Steve had done home construction and remodeling. So their agritourism selections reflect their own secondary ambitions, rather than being projects enforced onto them that they didn't want to get involved with in order to "save the farm." For another farm, these projects may be far from what the owners may want to be doing on their property. Others might prefer simple farm open houses for the general public that call for very few additions to the farm, or camp areas for bird watchers, or a multitude of different possibilities.

Year 1–The Fall Family Fun on the Farm began "by having a haunted house, straw bale maze and pumpkin patch. We served hot cider to our guests. This was the time for us to learn this new business of agritainment. We grew the business slowly. It is too much to comprehend at one time. Some of the things you try don't work as planned and you need time to correct them. It is a step-by-step and area-by-area process to grow your business. It is a time commitment and the mind and body need to adjust and adapt to all the different activities. We grew the agritainment business as follows:

Year 2–We accepted the challenge of designing something new every year to keep the people coming back. Our daughter, Jennie, 14, developed the theme for the haunted house "Who Murdered Dad?" She wrote the story about how the haunted house was a B&B and someone allegedly murdered Dad. These are guided tours through the haunted house, but actors have been place strategically to increase the anticipation of the guest as well as the tour guide adding suspense. That year, we planted six tenths of an acre of our yard into a football field size corn maze and coordinated its design with the murder theme and the solving of the mystery. We also had the straw bale maze.

Year 3–We added the miniature golf course. It is said to be one of the most difficult in the region. It is agricultural based with each "hole" based upon a crop or livestock enterprise in North Dakota. Obstacles are parts from machinery associated with the crop or livestock "featured" at the hole. The fairways and paths from one hole to the other feature the design of farm equipment such as a John Deere—a tractor pulling a dairy wagon. The golf course is very focused on educating the player about farming in North Dakota. Also, we added the "Tunnel of Doom" which is an optical illusion pathway in the dark that gives an effect of fast motion. This was used to create more interest to bring people back again. Teachers tell us this is a real "brain enhancer" which is used to stimulate the thought process of

children who have learning disabilities. We knew we were in the therapy/education family fun business, but not in the treatment of learning disabilities! The five-acre corn maze was about North Dakota, its road system and location of the 53 county seats. The haunted house was filled with scary stories. This was the year we started the school tours and added "Face Painting." Face Painting was done on children, teenagers and adults.

Year 4–We added the "Rat Racer" which is eight feet in diameter and four feet wide. It runs along a 150-foot track with rails and bumpers next to the golf course. This location of this attraction was to encourage guests to try the golf activity. This "rat racer" activity is really a challenge because the faster it moves, the faster you have to go. It is fun to watch children get themselves into a very fast mode. The 7.5-acre corn maze was about the United States. The continental 48 states were outlined and trivia about each state was presented. One entered and exited through the "International Peace Gardens." The haunted house theme was about a girl who inherited the family home. The pumpkin patch and straw bale maze continued to be part of the Fall Family Fun farm visit.

Year 5–We added "Tess," the Holstein cow. It was amazing that many of our school children in the rural state of North Dakota did not know that cheese was made from milk produced by cows. Children had great fun hand milking Tess. The older people who milked cows growing up said they would just as soon forget those bad memories. We really enjoyed teaching the children about agriculture, farming and a farmer's family work and life style. The haunted house theme was the journey of a young girl one stormy night. The eight-acre corn maze was of the human body. It was three-quarters of a mile around the body (skin). Everything was done in proportion so children had a real feel of where the lungs, heart, blood vessels, arms, legs and etc. were in relation to the whole body. It was a real educational reward to see the children respond to information about the body, health, and environment. The pumpkin patch continued to grow and the straw bale maze was continued.

Last year of diary—We are building a "Pumpkin House" this year. It is 12 feet high and shaped like a pumpkin with a stem on top. It will be used as a classroom inside to teach children how agricultural crops are grown and story time about animals, birds, and nature. It will also have shelves to display pumpkins for sale. We are making several improvements in landscaping around the buildings and display areas by using trellis and other structures. We are also planting other agricultural crops such as grapes. The haunted house theme this year is in the planning stages. The ten-acre corn maze this year is "the world." It will be 750 feet in diameter and will show latitude, longitude, airline routes to major cities and crops will be planted between continents. Since the sun moves 15 degrees each hour, time zones will be represented. We will still have Tess, the Pumpkin Patch, Straw Bale Maze, Face Painting, Golf, Rat Racer and The Tunnel of Doom.

Result—"We have grown to be a destination. Families will come from different cities in North Dakota, South Dakota and Minnesota to meet and spend a day at the farm. We have areas to relax, eat and enjoy the outdoors. Some families travel 300 miles to visit our farm. It is not unusual to have visitors travel 200 miles for a day visit and return home for work the next day

How they got their information and ideas: We do a lot of research and reading. The Internet is a great tool. We joined the North American Farmers Direct Marketing Association and have attended their trade shows since 1998. We listen to and meet people doing the same type of agritourism at conferences and trade shows. We discuss our ideas with the friends we have made across the country. We have also attended entertainment trade shows to look at ideas, talk to vendors and other entrepreneurs. We picked up the "Tunnel of Doom" idea at the Chicago Halloween Trade Show. Steve came home and designed and built it.

Enger Farm, www.engerfarm.com

– "Grow slowly and steadily: How one farm did it" is adapted with permission from United States Department of Agriculture, Natural Resources Conservation Service, "Alternative Farm Enterprises – Agritourism Success Stories," "Family Education and Entertainment on the Farm," May, 2002. ❧

Rules, Regulations & Liability Concerns 4

Rules, regulations and liability… oh, my! All businesses are surrounded by a multitude of lurking, possible regulations they may have to follow. No simple way to identify them exists, because each government entity is different and each business is different.

So, all we can do is start at the beginning and move forward. Begin by knowing what city, county or other jurisdiction your property falls under. If you're not certain, obtain its boundary identification from your city or county planning department. Know how your property is zoned (such as rural or rural intermediate.) Realize that in some rare cases, while your surrounding neighbors may be zoned rural, you could be something different. Be sure to know your own property's identification, location and zoning.

Then, go to your state or other geographical jurisdiction's government business website and work yourself through their checklist for operating a business in that area. One of the best sources for USA citizens in finding your own state's business rules website is through your Department of Agriculture website, which will often lead farming businesses to the necessary state pages. The website for Washington State's Department of Agriculture, for example, has pages covering Washington's rules and regulations for doing business in that state, as well as pages helpful to marketing, getting proper insurance, and much more.

Rules, regulations and liability of course, must be interpreted and handled locally with updated advice from your attorney, other professional advisors, local authorities, your insurance company and perhaps even your neighbors! However, if you engage in agritourism or directly sell products on your farm's premises, we can certainly cover some situations here that a hosting farm could come up against, discuss how other farms handled them, and point towards making the most of local and regional resources.

Zoning, permits and licenses

Greg Lawless of the Lawless Partnership in Seattle, Washington has many years of experience in the areas of business enterprises. In spite of his eye-catching last name (!), he has a great reputation for accuracy and good ethics. He says that one of the very first steps to take when planning an agritourism addition to your farm includes making sure your region or county allows the type of enterprise you're planning. "A property zoned agricultural and/or residential may not be zoned to allow some types of businesses to operate," he says. Although the American Farm Bureau Federation now recognizes agritourism as an on-farm activity, check first by calling your local planning department. Also, ask your planning department for a list of all the allowable uses for your property. If you're still in the planning stages, this list may be valuable for giving you additional ideas or ways you can modify a previous plan that you discover isn't in compliance with zoning laws. Whether your plans include a full B&B, roadside stand or other project, you may be pleasantly surprised to find that you're permitted "by right" to do all that you had planned without special permits, and a simple free- or low-cost application, if even that, is all that's necessary.

Or, you may find you need to make adjustments to remain legal. For example, you may discover that bed and breakfasts in your area have a limit of four bedrooms, when you'd planned for six. Take a look at the list of other allowable activities. Could those two extra rooms instead be legally used for a massage therapist in the family to give treatments (see Chapter 17), or become a retail outlet for farmed goods instead, both of which could complement your B&B operation?

Finalize your plan, then re-check it with your planning department. You may find you can begin tomorrow. Or you may be told you first have to make a way for more cars to park off the main road, or put a sign in a better location than you'd originally planned so it won't obstruct roadway views. Or you may find you do need a "use permit" for certain activities. If this is the case, get all the information you can on the use permit requirements from your local planning department, and seek qualified advisors and possibly insights from other farmers in your area who've been through the use permit process.

Liability insurance

"Accidents don't happen often, but they do happen and you want to be prepared," says John Ivanko, owner along with Lisa Kivirist of Inn Serendipity Farm and Bed & Breakfast, which provides visitors a farmstay experience that includes roaming chickens and fresh, seasonal home-cooked meals with fresh, organic ingredients from garden plots within a hundred feet of their back door. "Check with your insurance provider and see what they recommend," John says.

Horses are a huge draw for community and tourists, but insurance companies are cautious about them. Be sure to explain if you have extra safety features, such as horses behind fences and out of reach for distant viewing only, or if you provide extra one-on-one supervision.

Your insurance provider has to know exactly what your agritourism operation involves. In some cases, for example, free promotional farm tours or on-farm stores may require different policies than on-farm entertainment activities where customers pay to participate. The insurance company needs to know more than just a casual statement that some customers may occasionally come to your farm.

Some agritourism enterprise owners have had good experiences with their current insurance provider as the best place to start exploring how to get the liability coverage they need. You, however, may choose to look further if you want to compare

Members of a local Slow Food convivium gather for a harvest feast in an old apple orchard. For one-time special events, make sure your insurance company knows exactly what is going to happen, including but not limited to whether guests pay a fee to enter the farm event.

companies and packages, or if your current provider won't cover your new venture, or does not offer professional advice in this area.

If the latter happens, learn to persevere when seeking insurance coverage. An insurance rider—a one-time extension or addition to your usual insurance coverage—may be appropriate for the farm with just a one-time farm open-house or annual festival. One couple who was trying to get a one-day, one million-dollar event rider from their current insurance provider was told repeatedly by the provider that such a rider was impossible; they didn't offer it, and the couple would have to purchase a separate and very expensive alternative insurance product from them to receive the extra coverage they needed for that single event. Exasperated and about to give up on the event altogether, a friend offered to call their company and ask for the same thing, and within half an hour, the exact insurance rider the couple was looking for was in place and ended up costing them a one-time fee of $12. Remember that what appears to be the final word is not always the final word!

You may even want to consult your attorney. "I am biased, of course," Greg Lawless says, "but I think an attorney would be very helpful in all phases of (an agritourism) project." Greg says the attorney can help the agritourism project owner figure out what type of business entity to use (sole proprietorship, corporation, LLC, limited partnership, etc.— see Chapter 5); can help with any leasing issues of other vendors on the farm, and can help greatly with insurance protection. "The attorney can work with the insurance professional to make sure the client's expectations are met in terms of insurance coverage," Greg says. "Keep in mind, the attorney is just representing the client's interests; the insurance professional works for the insurance company."

Examples of what you might ask and learn from your attorney and other advisors on insurance include how and if disclaimers, waivers, on-farm safety signs, first aid kits, CPR training and other additions may or may not help protect you and your customers.

Also, your advisors may suggest safe and legal ways to save money on insurance and add to your protection at the same time. Tony and Carol Azevedo, owners of Double T Acres dairy ranch in Califor-

CHAPTER FOUR

nia, added weddings and other on-farm events to their business when guests who had come for a private family gathering enjoyed the ranch so much, they asked the Azevedos if they could hold their own events on the ranch. Tony and Carol reported in the USDA's Natural Resources Conservation Service's survey on successful alternative farm enterprises that at the time they began operation, they just needed a million-dollar policy, and they kept their homeowner's policy as well, which works for many farms that are also residences. But for some events, such as weddings held on Double T Acres, the Azevedos stated the client (such as the bride and groom) also has to provide additional insurance, which is usually a one-time rider (as described above) off their homeowner's policy for little or no extra cost to them. In Double T Acres' case, they stated that the client's insurance would pay first in the event of an accident, and any guests who got hurt would have to sue the bride and groom before they sued the farm. This gave their farm a buffer against sue-happy people, and encouraged the clients to be responsible during the event. That is just one example of hidden possibilities that may make your agritourism project financially viable when you once thought liability risk and cost made it too prohibitive. See also Chapter 5 on how your assets possibly can be further protected beyond liability insurance by choosing land trusts or a different business structure.

Finally, from the book, *Agritourism and Nature Tourism in California* from the University of California, here's an adapted list of further questions you or your attorney may want to include when consulting with your insurance company:

- Is there a deductible? If yes, how much?
- What does the insurance apply to?
- Premises and operations liability?
- Contractual liability to others?

- Personal injury liability to others including libel, slander, invasion of privacy?
- Advertising injury to others?
- Property liability damage to others?
- Incidental medical malpractice liability resulting from your helping an injured person?
- Non-owned watercraft liability?
- Host-liquor liability?
- Court costs for defense—above limit or included in liability policy limit?
- Also, is every employee additionally insured?
- Is the premium a set fee?
- Is the premium based on gross sales or on client days?
- Is association membership required to purchase this insurance, and if so, what is the cost?
- Does the insurance agent understand your proposed agritourism enterprise?

If you are shopping for new liability insurance for your agritourism project, a good place to start may be the insurance referral list offered to members of the North American Farmers' Direct Marketing Association (NAFDMA), www.nafdma.com. As mentioned above, now that the American Farm Bureau Federation recognizes agritourism as an on-farm activity, this may serve those trying to get farm-friendly insurance companies already covering their farms to cover this added activity. Also, other successful agritourism farms may have insurance companies to recommend. Finally, success with insurance companies surely comes easier with agents who are not ignorant of progressive farming professions, rather than those who rely on Hollywood stereotypes to understand the farming industry.

Searching for an attorney

Since rules and their interpretations become outdated, and human error happens in the publishing industry, and rules vary in different locations, never consider a book or article, either print or online, as your final legal nor financial advice for which you should seek a qualified professional. Many agree that the best attorney for final decision-making is one from your own location who has had experience in that location for at least a few years. As to how to find a lawyer, Greg Lawless recommends going to Martindale Hubble's web page which allows you to search by geographic area and topic:

http://www.lawyers.com.

Go to the "Business Users" tab, and under that find "Business Enterprises."

Help from your local agriculture extension agent

Besides your local state's Department of Agriculture, another source of often-free information is your local cooperative extension, which operates to help farms succeed and progress in all areas, including the interpretation of local rules and regulations. To find your local extension, go to the Cooperative State Research, Education, and Extension Service (CSREES) website (www.csrees.usda.gov).

According to the CSREES website, "Small businesses are a vital part of the U.S. economy, creating 80 percent of its jobs and accounting for nearly 40 percent of its private sales. Whether they are farmers exploring new agritourism opportunities, start-up companies in high technology, or micro-business entrepreneurs offering services from their homes, small and home-based businesses fuel economic growth in communities, including rural ones, where more traditional job opportunities are declining. Many of those communities are adjusting to changing market, environmental, and policy forces. CSREES' role is to supply science-based information and to help entrepreneurs understand these changes and take full advantage of emerging opportunities. Working with its many partners, CSREES promotes high-quality research to stimulate innovation in the private sector. It also helps build the skills and capacity of outreach professionals whose job it is to help entrepreneurs initiate and expand their businesses."

Through it all, keep the inspiration of your agribusiness project fueling you. There may be adjustments and delays, but many happy agritourism farmers finally have made it through the rules, regulations and liability desert to at last operate a business they now thoroughly enjoy. ❧

Choosing A Business Structure 5

Insurance isn't the only way to protect your assets. Nor are higher volume or prices the only way to raise your bottom line. You may want to consider the different business structures available to your agritourism farm, especially if you're bringing customers onto your property for the first time. The right business structure can put your farm into the category of other business owners who find their personal assets are much more secure, and their income much higher (sometimes by tens of thousands of dollars a year), just by switching to a new type of business entity.

It is not the intention of this chapter to give legal advice. The issue of choosing a business structure for your farm is rewarding, though complicated. Every single venture is different and calls for different actions, so it is the intent of this chapter to introduce possibilities some readers may not have considered, with the intent that you, the reader, consult with an attorney and do more research as only you can do for your particular situation and location. As with zoning and other regulations, remember, don't depend on a book (this one or any other) or article. What is right or best for your neighbor, or another very similar farm, could be discovered by the right advisors to be wrong for you.

The rules for the various business entities you may choose change periodically, and they are certainly different from region to region, but here, we'll introduce four that currently seem popular with various agribusiness farms: sole proprietorships, limited liability companies (LLCs), S corporations, and non-profit corporations. While the right choice can be very rewarding, the wrong choice, or the right choice misunderstood, can be devastating. There are other choices than the four listed here, including legal partnerships and various hybrids, and your professional advisors (including your attorney) can help you decide which is best for your circumstances. Some farmers, as you'll see below, even blend more than one business structure for a custom-made operating system.

A sole proprietorship is considered the simplest business structure in the USA. As a hypothetical example, Susan Smith grows heirloom peppers and tomatoes on her two acres and sells them to three local restaurants. She's named her business Susan's Heirlooms, and for a small fee, she applied for her "doing business as" (DBA) license under this name so she could deposit checks made out in her business name and have a business account at her bank. A sole proprietorship business structure is described as an extension of the person who owns the business. As a sole proprietorship, Susan's accounting can be more casual than for some other business structures, and she may be able to buy her groceries with the same checking account where she deposits her farm income. Most authorities consulted on sole proprietorship say, however, that if the business gets sued, so automatically does the owner (therefore risking

Susan's personal property along with any business-related property), as both Susan Smith, and Susan's Heirlooms, are one in the same.

Moving beyond a sole proprietorship. Susan now wants to add an agritourism element to her business. She plans to give paid tours of her organic growing operation, and offer garden-to-table cooking workshops on her property featuring tomatoes and peppers. She eventually decides she wants more asset protection even though she has adequate insurance. Her cousin showed interest in her expanding businesses, so she's also considering the idea of formal shareholders if she eventually builds a large, separate, certified kitchen where classes are held full time. Also, she has been told by her accountant that she is overpaying on taxes as a sole proprietorship, and could save money if she chooses a business structure other than a sole proprietorship.

At this point, she's decided to choose among an S corporation, a non-profit corporation for which she'd pay herself and others a salary, and a limited liability company (LLC), or a combination of the above. She has a lot to weigh and consider. Which structure or combination of structures would be best for tax purposes? For liability and protection of her personal assets when people consume her products and come onto her property? For image and marketing? And for a business lifestyle that suits her?

These more complex business structures can offer certain benefits that sole proprietorships don't offer, such as keeping your personal property and other assets separate and out of the picture if your business is sued. But the business must be operated properly for these benefits to apply in the case of a

Nursery items grown for sale on Tree Frog Farm, Inc.
www.treefrogfarm.com

lawsuit, tax savings, and other positive advantages. If Susan sets up an S corporation, for example, and doesn't treat the corporation like a corporation (such as forgetting to hold and record regular, formal meetings, or not keeping accounting completely separate from personal accounting), its protection may not be valid if audited or a lawsuit is filed. Also, people can simultaneously sue both a business and its owner personally, so Susan's assets may not be automatically protected by the business structure.

S Corporations

There are two types of for-profit corporations, a "C" corporation, and an "S" corporation. Both can help protect assets and save money for the business in ways no other structure can. Even so, the term "corporate agribusiness" is synonymous to "monster multi-national earth-destroying conglomerate" to some in the sustainable, local-farming arena. But many people don't realize that a corporation can be an ethical, locally grounded small family business, even owned by just one person. It does not have to

CHAPTER FIVE

be a huge centralized system set out to take over countries and consume the planet.

Unlike the sole proprietorship, the corporation becomes essentially a separate entity from its owner, wherein lies the reason there can be great tax savings and personal asset protection. When someone sues a properly operated corporation, for instance, the owner's personal assets are separate from the business, and therefore are not in jeopardy. But it is more complicated to operate than a sole proprietorship. As just one example, remember how Susan buys her groceries from the Susan's Heirlooms checking account? If you own a corporation, you cannot make personal purchases with–or even personal deposits into–the corporate account. This means that if your aunt gives you a check as a Christmas gift in your name, you can't just deposit it into the corporate account, where only business money can go in and out. You must pay yourself a pre-determined salary from your own corporation (a salary of which you can then do with what you please) and must always keep personal money separate from corporate money.

C corporations

C Corporations are more often used for very large businesses, while the S corporation is more often advised for smaller businesses, according to a SCORE counselor who helped with this chapter. John Robinson and Diana Pepper, owners of Tree Frog Farm, Inc., operate their one-third acre farm in Washington State and are very happy with their choice of an S corporation. They sell native plants, value-added body and health care products made from their own crops, and engage in agritourism by allowing visitors to tour their farm and come to their farm's yurt for classes, retail products, healing sessions with Diana, and other offerings. "Our accountant," says co-owner Diana Pepper, "suggested we incorporate. Because the corporation is separate from us–the individual owners–there are some tax advantages. We have been able to add money to the business as a loan that we pay back to ourselves instead of a capital contribution that stays in the business. The business rents the yurt from us for classes and my client work. Instead of taking vehicle mileage as a business expense on our federal income tax, we are able to reimburse ourselves for business mileage. Also, the business reimburses us for mortgage interest, insurance and taxes on the portion of the house that it uses. In this way we have been able to get cash reimbursements from the corporation outside of wages. As a sole proprietor and partner, we had to pay our full social security tax based on the business net income. When we pay ourselves wages as corporate employees, the business pays one-half of our social security tax based on our wages. Also, we are able to take stock dividends, which are taxed at a lower rate, in addition to wages."

Like many smaller businesses choosing the for-profit corporation structure, Tree Frog Farm owners were advised to operate as an S corporation versus C corporation. "For our size," says Diana, "(an S corporation) is less complicated."

But don't discount the C corporation automatically just because most smaller businesses operate as an S. Learn what its advantages might be to your particular farm as well before making a decision. "I believe it is easier to grow larger," says the SHARE counselor about the C corporation versus the S corporation, "and you can have an unlimited number of shareholders from around the world." Once again, the idea of multi-national corporations sounds like the ultimate bad guy, with shareholders from many nations owning a business that operates in a specific country. Good point. But consider. The Slow Food Movement that started in Italy is an incorporated entity with members from many different countries, and is instrumental in reviving sustainable local farms and local culture across the globe. Could this type of ethic ever transfer to other types of for-profit corporations? One of the foundations of the Slow Food Movement is that it insists its fundamental values are always interpreted and man-

Inn Serendipity Farm and Bed & Breakfast
www.innserendipity.com

aged at the local level in a manner that benefits that particular region.

Tree Frog Farm, Inc. is a single farm business offering several farm-related products and services. Other business owners have used the corporate structure as an umbrella under which to operate their farm along with other non-farming businesses. John Ivanko and Lisa Kivirist, co-authors of Rural Renaissance, and owners of their farm's bed and breakfast, Inn Serendipity, have also chosen an S corporation. "We formed a sub-chapter S corporation a long time ago," John says, "through which we operate our diversified businesses with an emphasis on hospitality businesses. Seems to keep things cleaner and better structured."

Your accountant and attorney should help make sure you understand how these business entities should be operated. Greg Lawless, the attorney from Seattle, Washington, suggests you have your attorney work along with your accountant when setting up an entity such as a corporation or a limited liability company (LLC, see below).

"I love to work with my client's accountant, and for most large transactions, insist on it," Greg says. "The CPA usually has the additional benefit of being familiar with the client's past tax returns, and capital depreciation schedules, so is in a better position to know the tax impact of a transaction. The lawyer is responsible for protecting the client from liability, identifying and properly drafting the necessary documents, and properly structuring the transaction that gives rise to setting up an LLC or corporation. Together they make a formidable team. The client might worry about it costing twice as much, but it really doesn't work that way. The CPA is going to be much more efficient with the tax matters, the

lawyer much more efficient at setting up the entity. For example, if you hired me to do everything, I would ask to see the client's tax returns for the last few years, need to analyze depreciation schedules, figure out future tax planning, etc. In other words, I would spend hours just getting to understand what the CPA already knows."

While it can be tempting to save money in the beginning without hiring an accountant or an attorney, or by hiring only one or the other, in the long run, many agree it pays to start out with both. "I'm really glad we had our accountant and attorney help us," Diana says about setting up their farm's corporation. "It was somewhat expensive to set up originally, but I think it has been worth it." However, there are resources for self-study and places to network that can give you a foundation of knowledge which will help you stay in the navigator's seat when setting up a business, and save money on questions you can have answered elsewhere, then have double-checked by your attorney. These are discussed below.

As far as the complications of the ongoing operation of a corporation, that depends on what you're

Quiet Creek Herb Farm, www.quietcreekherbfarm.com

CHAPTER FIVE

comfortable with. "The required corporate annual meeting and the annual corporate license renewal are no problem," Diana says. "Having all the tax advantages has helped us build a stronger business."

Non-Profit Corporations

Some farms even choose to operate at least part of their farm as a non-profit corporation. When done properly, this structure can allow certain grants open only to non-profits, and tax-deductible donations from corporations and private citizens, to help fund the farm tax exempt, and does not mean that the farmers don't get a salary and possibly certain expenses paid. Non-profit corporations have financial freedoms and obligations different than for-profit corporations. One of the most common complaints is the paperwork involved in setting them up. But it's well worth it to some.

Claire and Rusty Orner own the 30-acre Quiet Creek Herb Farm in Brookville, Pennsylvania, where they operate a 15-member CSA farm and raise vegetables, fruits, herbs, and edible- and cut-flowers. Much of their produce is turned into value-added products, such as herbal soaps, teas and vinegars. They wanted to add an educational agritourism element to their farm's package and so they turned their farm into a non-profit corporation: Quiet Creek Herb Farm and School of Country Living.

"The entire farm and school, including the CSA and farm store, now operate under the non-profit," says Claire. The farmland and buildings, however, are still owned by the Orners privately. "We lease the farm and buildings to the non-profit," she says. By turning Quiet Creek Farm into a non-profit, they were able to use grants and donations to add educational programs on the farm, and to then farm full-time instead of maintaining off-farm jobs to pay for the educational segment.

Pelindaba Lavender Farm, www.pelindabalavender.com

Limited Liability Companies

These can be composed of one or more individuals or business entities. The beautiful Pelindaba Lavender Farm for example, a destination eco-farm on an island in Washington State, operates under a limited liability company, Pelindaba Group, LLC. LLCs are the newest business structure, and a growing number of farms, including those involved in agritourism, operate as LLC entities. They are said to be less formal and complicated to operate than a corporation and can offer some, but not all, similar benefits as corporations, such as personal asset protection because they, too, become a separate entity from the owner. While they don't allow shareholders, they allow formal contributing members. Some LLC owners have reported that the initial set-up is more complicated and takes longer than for corporations. But Greg Lawless says that may not be the case anymore. "Each state has its own procedures, so it's hard to generalize," he says. Greg points out that in Washington State (for example), setting up an LLC is no harder than setting up a corporation. The filing fee for a corporation is $205.00; for an LLC $195.00, so even the fees charged by the State are comparable. Both those fees include Washington's "over the counter" charge, meaning a one-day turn-around for filing either entity. When LLCs first came into the picture as an entity, there was some uncertainty as to how they would work, versus corporations that have been around for a long time. But now most states have had them for some time and they are fairly commonplace. For example,

> *Many people don't realize that a corporation can be an ethical, locally grounded small family business, even owned by just one person. It does not have to be a huge centralized system set out to take over countries and consume the planet.*

Washington, one of the last states to adopt LLCs, has had them in existence since 1994.

How about blending business structures? While John Ivanko and Lisa Kivirist, owners of Inn Serendipity, put all of their various businesses under one S corporation umbrella, and Quiet Creek Farm put their entire farm enterprise into a single non-profit, some farmers choose to diversify into more than one business structure.

Greg Lawless states that when real estate is used commercially, the classic business situation is to put your commercial real estate into an LLC, put your business into an S corporation, and lease the real estate to the business. It gives the most liability protection, and best tax results. This could possibly work for a farm business depending on your circumstances, meaning the property itself is transferred to an LLC, but the farm business (e.g., Susan's Heirlooms) is made into an S corporation. But farmers' property is often also their own residence, which needs more consideration. The tax advantages of keeping your residence private rather than within an LLC need to be weighed. If your advisors feel this situation may be of benefit, make sure it's done with the help of an attorney who knows how to keep all your bases covered when it comes to transferring your property to your own LLC. As only a few examples (there are more), your attorney should check with your current mortgage lender to make sure your loan will transfer to an LLC without the lender calling your loan due, and should check that your title insurance remains intact.

Another business entity blend is demonstrated by Angelic Organics, a biodynamic Community Supported Agriculture farm in Illinois that started growing vegetables in 1990 and took on biodynamic growing practices and the CSA model in 1992. The owners eventually added an educational segment, The Angelic Organics Learning Center, to help spread the word on how their sustainable model makes farming profitable and enjoyable, and helps restore communities. While the main farm and CSA is operated as a for-profit sole proprietorship, the learning center is operated as a non-profit. They invite donors to give to the learning center on several attractive levels, including specific amounts needed to co-sponsor groups of at-risk youth for 10-week workshops, a beginning farmer for a 9-week, whole-farm business planning course, a low-income individual's attendance at a sustainable living workshop, and others.

Your own preliminary study

You will most likely operate with more confidence and save on CPA and attorney's fees if you first internalize at least a basic understanding of business structures before walking in cold. Your exploration may include your regional government's website, contacting volunteer business counselors, networking with other farmers who've been through the process, and reading an updated book on business entities written for the lay person. Remember that business structure rules change constantly and are interpreted differently by different authorities and regions, so make sure copyrights on the books you read are as recent as possible.

If you're in the USA, you can contact SCORE, www.score.org. SCORE, "Counselors to America's Small Business," is a treasure chest of free and confidential small business advice for entrepreneurs. Its counselors may be able to help when you're feeling overwhelmed or have precise questions during the planning stage but aren't yet ready for your attorney. Don't be intimidated if you feel that farming is too

Consider a land trust

While a land trust is not a business entity per se, it can be a valuable part of the overall package when considering the business entity under which you want to operate. The SCORE counselor for this chapter suggested looking into land trusts at the same time you're choosing business structures. In some cases, for example, property owners put their real estate into a land trust which, when done properly, can help create a protective bubble around the property in the case of liability concerns. The trust, somewhat like a business entity,

becomes its own separate legal entity in which property can be held for the benefit of one or more persons. If so desired, this trust also can be put into a business entity, such as a corporation, giving even more layers of protection.

As reported in The Progressive Farmer, there are both public and private land trusts. Public trusts are often those where farmers give or sell their land to a protective easement, such as American Farmland Trust, which preserves the land in the future as safe from

development. Private trusts, on the other hand, are more often for estate planning, helping to eliminate or defer estate taxes, and even creating guidelines for how the estate's assets are used over time to protect losses caused by spendthrift heirs. But these private trusts are also set up for protecting assets. If you were to cause an injury, for example, by bumping into someone at a public parking lot, that which is in a private trust (when created properly) can't be seized in a lawsuit by the injured party.

---

foreign a business for them. SCORE helped a Missouri dairy farm keep from closing down by adding an on-farm milk processor that bottled the milk in glass jars. The dairy farmer's business is now thriving, thanks to SCORE counselors.

Your farm and agribusiness colleagues also may be able to guide you through the processes somewhat. Understand that the ultimate decisions and accountability are yours, and those who turn to outdated library books, Internet articles, or friends of their hairdresser's uncle to get legal, business and financial advice often have horror stories to tell. But other successful agritourism farmers may be able to at least help you learn the basics in layman terms in

conjunction with the professional help you must seek. Rusty and Claire Orner of Quiet Creek Farm spent about a year and a half working towards their eventual non-profit status. But the Orners helped another farmer with his goal of reaching non-profit status, and his took six months less with their help. The Orners certainly used professionals to set up their operation, but have found that farmer-to-farmer networking in this way also is very valuable.

So, whether you end up feeling safer, saving thousands of dollars a year, or even find grant and donation money you never have to pay back to operate your farm, the world of business structures is a fascinating one well worth exploring. ✎

Breaking In: *The single farm festival or open house as a starting point* **6**

This chapter explores the idea of testing the waters of opening the farm to the public with a one-time, on-farm public event. If you're able to start slowly as described in Chapter 3, this chapter may be the one you turn to for your first public event after practicing with family and friends. But even if you've had practice, starting small at first with the general public can help you eventually grow larger with a stronger foundation beneath you to propel you forward. Here are a few examples of what a farm owner might be planning, and how to start gradually to build up experience and confidence.

Future plans to give walking or wagon farm tours to school children. Start by getting to know just one teacher; give a tour to his or her classroom only. Expand from there to all the children in that grade level, then to the whole school, then the entire local school district.

Future plans to offer large farm tour and open houses to the general public. Start by holding a picnic or potluck for family and friends, with the theme being educational and fun activities on the farm. Move on to offering a tour and open house to those you don't yet know, but are part of smaller and more controllable groups: the local senior center, garden club, a local Slow Food Convivium, or scout troupe and their parents. Organized groups such as these tend to be more self-governing and easier on crowd control, yet allow you to practice with an audience similar to the general public. When you feel confident, hold the event for the general public.

Future plans to host an annual themed festival. Start by being just a small part of another themed festival. Note any other community-wide festivals already taking place and drawing crowds, such as a fall art show, bicycle tour, birdwatching gathering, local food festival or home garden show. Contact the festival committee about adding your farm's open house as one of their larger event's many attractions, which means you don't have to "hold" the entire entertainment for everyone on your own the first year—you're just one of many. This also allows a larger entity, such as the birdwatching committee, to oversee the advertising and promotion of the event. By not being expected to be the sole entertainment for a crowd, you can test the waters with various activities: lavender wand demonstrations, composting lectures, etc., to see what you like and what visitors seem to enjoy. From this information, eventually hold your own event on its own, at a time that would be complementary, rather than competitive, for your town's other activities.

Or continue being part of your community's festival, growing larger and larger each year in a manner that complements this other event. For example, during organized bicycle tours of Skagit Valley Farmland, farm owners are encouraged to offer their farms as stop-off points and to put up

> *Continue being part of your community's festival, growing larger and larger each year in a manner that complements this other event. For example, during organized bicycle tours of Skagit Valley Farmland, farm owners are encouraged to offer their farms as stop-off points and to put up signs describing what crop is growing in their field.*

signs describing what crop is growing in their field. By participating, you are also adding to the bicycle tour's success, and can also offer brochures of your bed and breakfast, or your roadside stand or other attractions to generate future customers that pass by.

The 15-acre Sunflower Farm, Inc. in Georgia started their own Sunflower Festival after a group of neighboring antique tractor owners drove by the farm's flowering fields during their annual Fourth of July antique tractor parade. The view of the fields was so stunning, they felt that others should be allowed to enjoy the sight. So, the farm owners organized a weekend festival that coincided with the annual parade, both groups drawing visitors to each others' events. The Sunflower Festival is now an anticipated annual event for more than 10,000 visitors, with homegrown music, heritage artisans, local traditional foods, children's games and art projects, a U-cut sunflower field, acres of blooming field sunflowers, and the promise of the antique tractor parade going by each day of the festival.

Future plans to hold on-farm events with outside paying vendors. Start by organizing the festival with one or two other complementary farmers or product-related artisans in the area. Allow them to operate as a vendor for free, in trade for their partnership in preparing for the event and assessing its strengths and weaknesses afterwards as an outlet for other future vendors. Or, as a service to your community,

invite your first vendor to set up for free. When you aren't charging yet, the pressure is lower. Once you have more confidence and information to provide paying vendors, offer vending sites for a fee. Vendors will want to know how many people to expect, how much advertising you do, and that other non-competing vendors have enjoyed the experience in the past, in order to weigh the price you propose to charge them.

Planning the event

Walk yourself through the following six steps to plan and prepare for your event.

1. *Be clear on our specific goals for your event.*

Though the goals may seem obvious, you may be surprised and gain a tighter focus on your event if you really think this one through. What are your reasons for holding the event on the farm, and are any others involved aligned with your goals? Look over the following to see what comes to mind.

- ⬥ Make money directly at the event.
- ⬥ Generate farm promotion for other non-event farm product and service sales, such as visitors seeing your tulips in bloom, and therefore ordering the bulbs from you online, or touring your rare quince orchard as the fruits are ripening, and then noticing your preserved quince for sale in food co-ops.
- ⬥ Increase the pleasure of the farming family and community—it just sounds fun and is a good way to give back to the local community.
- ⬥ Contribute to the local economy by drawing tourists to the area seeking local cultural, historical or eco-tourism events.
- ⬥ Contribute to education about farming and to generate support for local farms from today and tomorrow's future adults.

Your final focused goal will no doubt be a combination of several of the above, allowing you, for example, to increase your bottom line in a manner that gives back to your local community's economy.

2. *Devise a theme and title for the event*

Knowing your goals can help you settle on a theme for the event, and its title, such as Sunflower Farm, Inc.'s "Sunflower Festival" described above.

The theme may be simply the offer to experience your particular farm. "Come to Foxhill Farm's Open House." The simple open house can work as a theme for farms who don't want to put a lot of extra time and effort into agritourism additions. Farming itself is a novelty to many people who will enjoy touring your dairy or watching how you distill your lavender to extract its essential oil. Success will depend on the number of interested citizens you can attract, and whether your farm is unique within its location rather than being surrounded by many other similar farms that also host agritourism events. "Open House" may not be enough of a theme statement for people to even know what's being grown on your farm and if it would interest them. "Foxhill Farm's Rare Wool Sheep Ranch Open House," would work better. Your remaining promotion, such as what's listed on a poster to be put up around town, would list: seeing rare breeds of sheep, touring the barns, watching a sheering demonstration, etc.

- May be related to the season, "Celebrate Spring at Foxhill Farm."
- May be related to a main crop such as a sweet corn or heirloom tomato tasting event, or "Corn Tasting Festival."
- May be related to events in the local community, such as a community local food festival. If the local food festival is named "Taste of Jonesville," add a trip to your farm to its schedule offering antique apples and goods baked from them, and name your event, "Taste of Jonesville's Antique Apple Farm Tour."
- May relate to a secular or spiritual holiday, such as Arbor Day, Christmas, Halloween/Harvest or

Harpist being invited to play during Mother's Day at a rhododendrum farm.

Celtic May Day: "Hollyhock Farm's Haunted Harvest Festival."

- May be related to a shared interest such as a celebration and demonstration of sustainable living, or bird watching. If your farm has a straw bale barn, uses solar energy, or other such innovations, sustainable living could be used as your theme along with your sustainably grown products: "Hilltop Farm's Sustainable Living Expo." If you'd like bird watching to be your theme, check with an expert from the local Audubon Society to see what species are prevalent in your area. "Goldfinch Days: Birdwatcher's Open House at Ferncrest Farm."
- May combine the last two, such as giving tours of your sustainable farm near Earth Day. "Earth Day at Ben & Bailey's Organic Produce Farm."

3. *Choose the date for the event*

Your goals and theme will help you choose the date for the event.

Will it depend on a specific crop's ripening time? Some annual events that depend on crop ripening will have different dates each year, but they will be similar from year to year as you predict just when the tulips will bloom or the blueberries will be ripe for U-pickers during your blueberry festival.

Is it seasonal? You may have more leeway in this case. Find out what weekend in the season doesn't conflict with other events in your area.

As mentioned above, do you specifically want to avoid other local conflicting festivals or national holidays that might take visitors away from your event? "I would recommend paying special attention to the dates of other activities in your area in order to avoid overlapping or conflicting with other established events," advises Bill Cummins, president of the Skagit Valley Farm Tour (see Chapter 11). Check for both local and national events. For example, you could easily lower your numbers if you have a family event during the World Series.

Check local weather patterns, also. Which weekend historically has the best weather? There's no guarantee against rain, but you can lessen the odds.

Is your event, instead, just the opposite, to be held in conjunction with another local festival or national holiday? As mentioned in Chapter 2, a rhododendron farm holds a Mothers' Day open house of their rhododendron forest that displays the full bloom of the potted rhodies and azaleas they sell. Many varieties are in blossom during Mothers' Day. They hire a musician (often a harpist) and employ food servers. It makes a wonderful place for families to take Mom on her special day. In this case, the national holiday enhances attendance, and obviously dictates the date of your event, but look ahead if you hope to have an annual event that repeats itself year after year. Are the farm and your family guaranteed to be available for this event each Mother's Day, or Easter, which has a date that changes every year?

4. *Vendors*

Knowing the goals, theme and dates will help you find paying vendors if and when that time comes. Note the wording on other farms' vendor applications in your state (find them online) with waivers written to keep their farm safe from certain liabilities that come with hosting paid vendors, and ask your insurance agent, attorney, or other advisors if this wording is the best you can do for your own event, and if they advise any other precautions. When choosing vendors, consider covering the following basics listed below (items in parenthesis show ways to think out of the box.) Your festival customers aren't going to want to see the same burger and fries vendors they see all summer long at other events.

Food vendors. They should complement your farm's products. A franchised ice cream vendor obviously wouldn't complement a farm promoting free-range dairy or its vegan-oriented produce. (Personal chefs are hired by families to cook regularly or for in-home parties, and they are often seeking ways to find new customers. Are there any local personal chefs who'd love to expose their business to your audience and who would also use fresh ingredients from your farm?)

Music. Check the music department of your local high school district or community college to see if a promising violinist or vocal quartet would like to set up and play for the experience of playing for a casual audience. Be sure to work with their teacher to make sure music pieces played are appropriate.

Other Entertainment for both adults and children can include vendors for cooking demonstrations, magicians, nature and farm demonstrations, animal viewing and petting, craft demonstrations. If you don't have farm animals of your own, see if there are those who travel with therapy pets willing to participate. Therapy animals are often dogs, but also llamas, horses, rabbits and others. See if a local children's birthday party event planner would like to set up as a vendor with a nature-themed outdoor children's fun area where they also sell their products and services.

Health or healing experiences. Beyond the healing of healthful food, music and contact with animals, you may also want to set up a chair massage corner or allow an aromatherapist to offer short treatment demonstrations One farm enjoys a "healing temple" at their annual gathering, set up under a canopy, allowing a variety of healing modalities to be demonstrated.

Craft and art items for sale. Summer art shows that many towns host can start to look alike, with seemingly the same pottery and whirligigs sold year after year. Ask your local high school or community college art director if there are any very unique artisans in your area that don't generally follow the art show circuit. Look for locally produced soaps, syrups or other goods that would complement your own products and interest the potential visitors to your farm. Make sure amateurs just breaking in have an ample supply to sell to your guests.

5. *Organizing festival help*

Knowing the theme, date, goals and other vendors (if any) of the festival helps determine who needs to do what during the festival, and who else should be solicited to help out. You may need to determine this simultaneously with #6 below.

Start with farm owners/partners. Who wants to handle the promotion (see Chapters 7 – 11 for promotion and marketing)? The budget? (You'll need to know how every dime is spent on this project, including any added liability insurance, and assess its direct and indirect financial value afterwards.) Talking to people once they come to the farm? Operate the entrance gateway? Handle the CPR and first aid area? (See Chapter 17 for more safety and first aid information.)

Will you need more volunteers or paid help? How many and for what?

Will close friends, neighbors or extended family be taking any positions?

Depending on your plan, would any of the following be sources of further promotion and/or volunteer help, such as offering to hand out or display fliers of your event: Local schools; churches; Chamber of Commerce; local tourism board; local social,

This lavender farm sets out unique attractions during its lavender festival, including this outdoor place to play chess.

Pelindaba Lavender Farm in the Pacific Northwest, USA, puts up extra signs to guide visitors during their annual lavender festival.

shared-interest, and civic organizations; scout troups looking for wholesome community service activities; or a local Slow Food Convivium. If it turns out you need more actual help at the festival than you have from within your own circle of free help, and can't yet afford to hire help, ask for volunteers from your church, CSA membership, or local scout troup, letting them know the festival helps people learn about their food supply and helps farmers continue to earn a living from farming. Offer them some form of farm-related gratitude gift.

What methods will you utilize to assess attendee feedback and/or solicit their suggestions for future improvements to the event (see Chapter 3)?

6. *Now you're ready to create more detailed plans for setting up the festival:*

- Where will extra vehicles park?
- If there is a possibility that large numbers of bicyclists will come by, is there a safe place for them to chain up their bikes?
- Do you need to set up a specific entrance gateway and block off other entrances to the farm?
- Set out off-farm signs leading to the festival.

- Set up signs leading to bathrooms, washing areas and first aid/CPR area.
- Any off-limit areas need marking off?
- Set aside time before the event to put household valuables locked and away.
- Curb appeal: When will last minute garbage, laundry and grouchy dogs be put out of site?
- Set up a clean-up plan. Where should containers for trash be located?

7. *Afterwards*

Chances are, when the festival is finished, you'll find yourself floating in a cloud of (hopefully) exhilarated exhaustion, if not outright shock, and may feel pressured to jump back into the regular business of farming without assessing the success of the event while it's fresh in your mind. Be sure to plan a restful but productive "afterwards" gathering with those closest involved shortly after the event. Specifically set aside a date ahead of time just for this to assess the strengths and weaknesses of the event before it slips into oblivion, as well as study customer/attendee feedback with a view toward future improvements. As time passes, the impact of the event will continue to incubate, and new insights will surface. But don't wait too long, or worse, do nothing. Just know ahead of time you'll assess the event shortly afterwards, using fresh event memories and statistics to tweak your farm agritourism business plan, but that it may be further tweaked as answers to problems and new customers generated from the event become more clear. Many of its benefits, you'll no doubt find, will show up further down the road. But the festival still has to hold its own right away enough to make it feasible for the farm owners. ❧

Part III: Promotion

Promoting the Farm: 7
An overview of possibilities

This chapter is both for those who are just starting their exploration of farm promotion and wanting to survey the territory, or those seasoned and looking to see if there are any gaps in their current system. It overviews the most popular basics of farm promotion, whether you're looking to draw attention to a long-standing roadside stand, or introducing a full-fledged agritourism service such as farm schools, Bed & Breakfasts, or children's tours of the farm. Many of the methods listed below are detailed in greater length in later chapters.

Be aware of both new and returning customers. Do you currently have a waiting list for your farm-stay B&B and prefer to concentrate on keeping the good customers you have returning? Or do tourist reservations seem to be dwindling? When choosing promotional techniques, consider that you'll most likely combine the need to attract new customers and keep past ones loyal and interested in your farm. Many promotional techniques will generate both repeat and new customers, while some are more appropriate for one than the other. To help yourself choose which ones to use, know which type of customers you're looking for.

Attracting the media

Numerous successful farms that cater to the local community insist that articles in local papers do more to draw quality customers than any advertisement one could possibly pay for. When you place an advertisement to say your farm is a fascinating place for children to visit, consumers know this endorsement is something you've paid for, but when a media person extols the fascination of your farm, readers are more likely to listen! Media coverage can also make current loyal customers proud to have already discovered you. My own experience with media coverage is that when they spell your name wrong, don't quite quote you correctly, and mistake goslings for ducklings, it still works! People love to witness what the media has reported. You may be discovered by the media in a variety of ways, but the best way is usually by sending out news releases to your local newspapers, TV stations and radio stations. This alerts the media to your existence. Even farms trying to reach a more distant audience can benefit from local media attention because little mentions can lead to larger ones. For example, larger regional or national publications sometimes get story leads by scanning various local papers. One destination farmer found his farm listed in European travel guides with no idea how his farm was initially discovered, but now happily books numerous Europeans through August by January or February.

The farm open house or one-time festival

A one-time or occasional on-farm event is sometimes used as the agritourism project itself, but it's also an effective tool for attracting the media and generating word-of-mouth promotion (very valuable). "We can't say enough about open houses and farm days," say John Ivanko and Lisa Kivirist, co-authors of Rural Renaissance and owners of Inn Serendipity, a bed and breakfast farmstay experience in Wisconsin, USA. "It's a great opportunity to form relationships with current and future customers, or in our case, B&B guests. We have one near July 4th, celebrating 'energy independence day.'"

Successful open houses have been as simple as the farmer stacking a table with ears of various heirloom sweet corn varieties next to a pot of boiling water, with salt and butter nearby, for a corn tasting event. The event itself, of course, will need promotion to attract visitors, but once the event occurs, it may end up being the most valuable promotional tool of all. A news release should be sent out to all local media, alerting them to your event.

Fliers

For one-time or seasonal events, some farmers have benefited from printing up fliers to display in the towns where their customers live. The owners of Sweet Grass Dairy in Thomasville, Georgia USA hold two open houses in spring and fall called Market Day that attracts many visitors to their grass-fed dairy. Co-owner Jessica Little states that one of their methods for drawing customers includes having their local printer create posters announcing the event, which they put up around town before Market Day.

If interested in trying out fliers or posters, get to know all the locations near you that allow quality fliers or posters to be displayed, and make sure yours are likewise top quality—either by paying for, or

bartering for, the services of an excellent graphic designer, or finding an artistic friend to help you. Also, make sure the basics below are covered. I've seen fliers that herald a fantastic event, with no address listed and no contact information for finding out where it is!

Include:

- A short catch phrase in large, bold, easy-to-read letters (don't get wispy or flowery with the lettering here). You should be able to read it 10 – 15 feet away.

- In smaller lettering, list all other relevant information: date, time, fees, location, contact for more information. Use very few words for this list. Casual passersby will be your flier's audience, and may keep on walking if they see too many words. This list is most effective and attractive to the eye if bulleted.

- An attractive image of the farm, of flowers, of pumpkins. This can be almost as important, if not more so, than the catch phrase. It also should be visible from a distance.

- If you have a farm logo, or perhaps a logo just for this event, it should be on the flier in an artistic manner that doesn't make the flier look cluttered.

- Have a good speller (someone else than the designer) do a final proofreading.

- Generally, put it up two weeks before the event. One week may not give people enough time to plan.

- Remove fliers when the event is over if you want those kind enough to let you use their space for free advertising to let you put up fliers again in the future.

Brochures and business cards

Printed brochures are giving way to online promotion as a tree-friendly and more valuable form of ongoing marketing. But many still like to carry business cards, and you may choose to overlap and use both brochures plus business cards along with online marketing, making sure your website's URL is on your brochure and business cards. The brochure tends to be used as something adequate for describing the farm year-round, whereas the flier described above is for announcing a one-time situation. Check your Chamber of Commerce, church, or other organizations you're involved with to see about a permanent place to offer your brochures. Their occasional uses might include any off-farm community events you take part in, such as a community festival where you sell farmed products at a booth.

Online marketing

Jane Eckert, of Eckert AgriMarketing in St. Louis, Missouri, states that online marketing is of utmost importance to farm marketing today. There are always exceptions, as with the occasional farmer who finds a niche and doesn't even own a telephone. But most will want to consider the advantages of a website. Jane points out that approximately 75-percent of North Americans use the Internet to plan their family and business trips; to find entertainment and family experiences; and to purchase fresh farm products. "Online marketing opens new doors for farms, and customers are eager to shop this way," Jane says. "Even if farmers perhaps aren't familiar with the Internet, they must recognize that most of their customers are—and today's consumers rely on the Internet as their primary source of information." The basic need is a website that's kept fresh and up to date. Start by putting up your basic website (or adding your agritourism page if you already have a general farm website). Later, added elements can include the online farm journal via a weblog which helps retain current customers (see below), and various other methods for helping potential new customers discover your website such as trading links with other quality farm or agritourism websites, and finding free listings on agritourism and festival and event sites. (Note: when trading links, do so only with websites that are relevant and of interest to your audience—too much irrelevant, "junk" links

When you place an advertisement to say your farm is a fascinating place for children to visit, consumers know this endorsement is something you've paid for, but when a media person extols the fascination of your farm, readers are more likely to listen!

from your website will act as a negative on search engine ratings, and these ratings are what determines how quickly your site will be found by Internet users).

Signs and curb appeal

A farm sign can alert people to your location, create a gateway between the farming and non-farming world, and draw the attention and curiosity of passersby. Farms are often private homes, and visitors feel they are welcomed rather than intruding when a sign officially announces the farm. If you give the occasional, casual farm tour, an A-frame or otherwise portable sign may suit your needs. If drawing the attention of the general public is inadvisable for your privacy and safety, you may choose to forego an official, permanent farm sign. But curb appeal can be added instead of, and certainly in addition to, a permanent farm sign. In improving curb appeal, you'll want to maintain being true to yourself—after all, you're a farm owner, not a theme park. But perhaps tweaking your farm's first impression somewhat, such as deciding that the trash really could be moved to a better location than your front yard, won't make the farm any less functional. We get used to our own surroundings and don't get offended when we're shoving dinner dishes to the side to make space on the table. But for curb appeal, try to sort out what visible aspects are actually successful farming operation: neatly stacked buckets, farm wagons; and what is residual work and trash we all wish would go away but never does: piles of unsort-

Ride elevators... or at least think out of the box

Gift baskets are value-added products some farmers are selling both direct from the farm and off. Here's one person's secret to getting urban folks to order her gift baskets. She built a beautiful basket and carried it to a busy high rise's elevator, as though she were delivering the basket herself to someone in the building. All afternoon, she rode up and down the elevator, handing out business cards to the people who couldn't help being enthralled by the beautiful basket. This is how she got her first initial customers, and her business grew from there.

While there are definite, cookie-cutter recipes for marketing, many of which we've overviewed here, remember to think out of the box when it comes to generating promotion for your farm, and custom-fit it for the lifestyle you already live. Hate elevators? (So do I). Instead, do your kids belong to a scout troup? Could the troup and its parents have a VIP tour with a sample product and farm brochure? Do you knit? Could a knitting gathering take place on the farm? Are you involved in the local Audubon Society? Could they come to the farm

for bird watching? Do your siblings or parents have an upcoming anniversary?

Tony and Carol Azevedo of Double T Acres in California held a 500 person anniversary party for their parents on their dairy ranch. Afterwards, guests began calling to see if their own events could be held there, and since, they have earned substantial additional income from weddings and other events held on the farm.

ed junk mail, tossed buckets instead of stacked ones you swear you'll get to tomorrow, and the trash itself. Consider keeping the latter out of view.

Networking with larger organizations

It can be profitable to network with larger organizations created to help promote ecotourism. It can also be risky, or at least, a waste of money.

"Larger organizations" include citizen groups specifically formed to promote and support local farms, local Chambers of Commerce, agriculture extension groups, shared interest groups such as sustainable living organizations, and businesses set up to lead tourists to attractions by geographical area.

Networking organizations also may include those with which you create packages with other businesses such as hotels. Or it could mean looking into all other events that already happen in your area and joining them, allowing your farm to be a seg-

ment of those events, such as being a stop-off point for an annual bicycling event.

Tourist promotional groups and listing directories may charge a fee to list you, or may want a percentage of your revenue, depending on your locale and theirs.

The main thing to watch out for is whether the entity really wants to make their relationship with you mutually beneficial, or simply wants to take an excessive or unfair percentage off the top of the fees you charge guests without doing much authentic promotion beyond a quick directory listing. You are stuck doing most of the promotion while they take their slice. In some countries, there is the problem of local farms being promoted as tourist attractions by large tourist organizations, without any method for the farmers themselves to receive added revenue. Also, be aware of restrictions that are too high for you to meet. Bed and breakfast promotional organizations, for example, understandably need to make sure the B&Bs they endorse are safe and that the

B&B is operated legally, in order to maintain their own reputations, which allows them to stay successfully in business. But their restrictions may be too stiff for your perfectly legal and enjoyable B&B farmstay. If they require private bathrooms and you offer a rustic outhouse, you'll find that marketing independently will serve you better.

The farm journal

The farmer's own unique story is another of the greatest marketing tools available. Both men and women farmers of the past used to keep journals and logs; it's a natural part of farming for many. If it is for you, consider using a version that's appropriate for others to read as a marketing tool. Today, the farm journal meant for others to read can be given out in the form of a monthly, printed "Letter from the Farm," an online "Farm Diary," or a "Journal of Greenbrook Farm" sent electronically via a variety of ever-evolving new methods (see Chapter 8 on methods for sending messages out electronically.) Numerous farmers have found monthly newsletters to be very supportive. The farm journal can be a large part of this newsletter, or a separate piece of literature offered at the current or potential customer's option. With such a journal, it's the realness and the uniqueness of who you are that cannot be duplicated by theme parks or slick ads, and cannot be mirrored and offered for less by a competitor. No one can describe the first crops that push up from the soil, complain about the unexpected hail storm, share excitement about the new peach variety, the way you can. Let your guests live vicariously through you. A farm journal can maintain current customer loyalty and attract new customers when its print version is passed around, or if it's an online version that can be found by web surfers.

Paid advertising

There are books, trade magazines, in fact entire university degrees catering to the benefits, pitfalls and skills needed to make paid advertising in print, radio and TV media effective. With so much already available, we won't elaborate too much in this book. But do know that the reason news articles, for example, have more draw to customers than paid advertising is their believability and trust. The media is supposed to give unbiased reports and is therefore more trusted by the public. Ads, most people know, are meant to say only the good things, and leave out the rest. People don't trust paid ads as much as they do other forms of communication such as news articles and going in person to the farm.

However, a classy paid advertisement can sometimes pay off. There are two basic types in print: Display ads, or space ads, take up space, say four inches by four inches on a newspaper page, or the full back page of a magazine. The other choice is the classified ads.

If you are trying to draw customers of a certain location or shared interest group on an ongoing basis, and you have a reputable publication that reaches this audience, a simple, repeated space or classified ad may help keep your farm's name in the public eye. If you're having a one-time farm event, and you feel that it may appeal to those who love to cruise classified ads (bargain hunters or those seeking special events), a classified ad may be well worth its cost. Display ads, too, may work for a one-time or special event. Study ads from previous editions of the magazine or newspaper and find out which categories for classified ads ("pick-your-own," "family events," etc.), or pages or sections for display ads, are appropriate for what you're offering—chances are, that's where the customers will be looking.

If you plan to get deep into paid advertising, I strongly suggest you learn more about how to target such ads, and how to keep track of their effectiveness over time. Also, learn to talk shop with fast-talking advertising salespeople. For example, newspapers can put your ads in the least desirable location. Ask what your options are as far as where your ad will be placed, and if you have no control over where it's put, and there's a chance it will be stuck in a rarely-read location, or a location that doesn't draw the type of reader who would be interested in your farm, you may want to tell the salesperson you'd rather do other forms of promotion than risk purchasing space from his or her newspaper.

As another example, if a reporter is doing an article related to your farm's theme, perhaps even an article on your farm itself, you may be suddenly approached by the paper's sales staff, with stiff deadlines, to place an ad at the same time the article will appear. They will tell you that, since there's going to be an article about your farm, now would be a very effective time to purchase an ad in the paper. They may be correct, but consider this very carefully. In what way would a paid advertisement benefit the promotion you'll already receive from the article? Would it actually look better for you to be presented only in article format? Would your advertising budget do better to allow the article to promote you for free, with, perhaps, a follow-up ad in the next edition? If you feel you'd rather not purchase an ad on top of the news already being presented about you, you may want to tell the salesperson that your advertising budget has already been allocated for the year, and if the article about you appears to attract a good enough audience, you'll consider them in your budget next year after carefully, without pressure, assessing their information regarding readership numbers, advertising prices, and choices for ad location.

Media employees, on the other hand—if they are honest and experienced—often can be your best advisors as to size and placement of ads. They are in a position to know what has worked for other advertisers. You can also check references by calling a few people who have placed similar ads (as long as their ads are not too competitive to yours) and ask for their feedback—how their ads are working for them. Also study ads for a period of time to see if ads are repeated—a good indication they're working.

Again, selective and especially low-cost advertising such as classifieds can often be effective for small businesses, but especially in the beginning, explore all the free or low-cost promotion available before pouring money down the "advertising hole."

Promoting with positive customer feedback

Customer feedback should do two things to boost your business. It should reveal what makes your customers happy, so you know what to continue. Or, it should tell you where you really screwed up. The

Unique art graces the lavender fields of Pelindaba Lavender Farm, while portable pavillions are seen in the background with various activities for visitors at their popular agritourism event, the yearly lavender festival.

Promotional help close to home

Cooperative State Research, Education, and Extension Service (CSREES) offers a state-by-state resource list of private and public organizations that help small farms with their projects, including the marketing of on-farm and agritourism enterprises. Search for their Small Farm Resource Guide:

www.csrees.usda.gov

According to CSREES website: "CSREES funds and supports research, higher education, and extension activities related to marketing and trade. In addition, the agency partners with public and private sector organizations to promote successful marketing and trade methods and strategies including alternative markets, products, policies, and institutions.

"Agriculture—in the broadest sense of the term—is in the midst of a major revolution that will change how food and fiber are produced, processed, distributed, and marketed in the U.S. and abroad. This has a significant impact on farm and ranch families and firms, agribusinesses, and rural communities – requiring them to make major strategic decisions to be successful… Marketing has become a crucial element of contemporary CSREES and land-grant university work."

Word of mouth promotion for farmers and market gardeners—a free download available from New World Publishing's website at:

www.nwpub.net.

---

six-foot tall, scary monster that chased the preschoolers out the door in the haunted barn was just plain stupid. Use the negative feedback to tweak your product (that makes you smart again), such as changing the monster to friendly "Mr. Pumpkin" or finding a different audience for the original monster (such as teenagers who love to be scared by six-foot, scary things).

Your customers' positive feelings also work for you by generating some of the best promotion money can't buy: word-of-mouth. Further, you can use positive feedback as another way for marketing: Build a reliable group of satisfied farm visitors for references or testimonials. This foundation is the leverage many small businesses use to great advantage.

Word-of-mouth promotion, in fact, can be used even before you've decided exactly what your direct marketing venture will be. New business owners sometimes use a strategy where they contact potential customers even before any product is created. A list of possible customers is drawn up. A detailed description of the potential product or service is written, and then it's sent out to the possible custom-

ers for their feedback. The reasoning is that most of these hypothetical "customers" will give both positive and negative feedback. With the negative feedback, the product is improved, and the description or prototype is sent out again until a desirable "can't miss" product or service is developed. After this, a few of the products are given away as a practice run to the most eager customers who receive it in trade for giving testimonials and references. This allows the new business to avoid wasting money on a product that no one wanted in the first place, and to start out with a group of already very-satisfied customers. They start with a proven product with satisfied customers to push them off towards success, using those first testimonials on their websites, in their brochures, and anywhere else testimonials are appropriate.

This has worked especially well for new agritourism ventures when the farms already had a list of potential customers, such as CSA members, and the new agritourism project, such as an annual pumpkin patch, is in the planning stage. The farmer already has a list of potential customers from which to get feedback and to practice on in trade for testimonials.

There is one flaw in the idea of letting customers have too much say in what your on-farm project will be, and this is discussed further in Chapter 3 in the section about helping your agritourism project progress and improve with ongoing customer feedback. Some people are so removed from nature and the satisfaction of hands-on creation and long-term commitments today, they may not even know how much they'd enjoy certain on-farm activities until they experience them in person. Your farm's road-side flower stand can entice them, but once there, they have to actually see the farm's activities to realize how much they want to take part in them. So, for a growing number of people, it's you, and your unique passions and atmospheres those passions create, that they are seeking but don't know it until they see it, and you need to find this type of audience rather than the demanding, instant gratification "entertain me," audience. Don't let the latter dictate your farm's direction with their feedback. This group wants fast stimulation, and gets bored easily. The former wants depth and authenticity. ❧

Online Marketing   8

It is said that the Internet reinvents itself every three months or sooner. But once you find yourself using your computer and the Internet with at least the current basics as a marketing tool, the upgrades and progress usually come as easy to implement as new information on farming methods. Just being involved in it hands-on makes it easier to understand and move forward with.

Charley and Ginny Hein, owners of Charley's Farm in Washington State, had a dream to start a small farming venture, and chose to grow rare, gourmet garlic. In 1997, they began with four varieties. Today, they grow 30. Customers can come right onto the farm to pick up their garlic or they can have it shipped. Either way, the Hein's computer has been a major marketing tool.

"We can't afford to sell wholesale to other businesses," Ginny says, "so, all of our sales have been either word of mouth or through the Internet. We sell out our garlic crop each year this way."

Most farmers start with a website as their door to online marketing. Some farmers have pointed out that a website is like having a multi-page, full-color brochure that costs pennies rather than the thousands of dollars it costs for a printed brochure—with the advantage, also, that it can be easily and inexpensively updated.

Jane Eckert of Eckert Agrimarketing travels the country helping farmers market their products, and says that every farm should have a website to meet the needs of their customers. "A website is an absolute necessity," she says. "For direct farm retail businesses, not having a website is like not having a telephone! Actually, I believe a website is now a more important marketing tool than a brochure."

Along with a website, some farmers add an electronic newsletter and a weblog or "blog" to their online marketing package. The e-newsletter (or " e-zine") is the electronic version of the monthly (or so) print newsletter many farms, especially community supported agriculture farms, occasionally tuck into their customers' shares. Farmers report their newsletters have invaluable customer appeal. Jane pointed out that e-newsletters offer on-going communications on the farm's progress to their best customers who can then find reasons to return.

Blogs are sometimes thought of as online farm reports of customer on-farm activities or ripening times that guests can open, read and even add to with their own comments. Little bits and pieces written two or three times a week by the farmer supplement or replace the e-newsletter—and some say are the hottest thing in online marketing. Unlike the e-newsletter, which is a one-way communication from the farm, when customers are allowed to add their own comments to blogs, valuable marketing information about customers' desires can be gleaned.

Beyond the basic website, e-newsletter, and blog, there is a fourth way to draw customers online. Bill Gluth owns "Develop Your Vision," which is a home-based business consulting service operating out of Surprise, Arizona that helps people in many types of businesses succeed. He already had his own published website, and was looking for ways to generate more visitors to the site, and ultimately more customers. Then he hit on the idea of adding three more things to his website: the e-newsletter and weblog already described, but also, free, short, educational online articles, or "ezine articles" which are short pieces written and published on various online sites visited by many people that otherwise wouldn't know about your farm. The ezine articles are written about topics that might interest the type of customer you're seeking, and they also have a link that leads to your farm's website. When the potential customer finds that article and reads it, he or she may be enticed to the farm's website and discovers its products and services. Bill has called this combination of website, e-newsletter, blog and ezine articles "edutising," and has carefully analyzed its benefits, and says it has brought him and others he's consulted added revenue and new clients. He says that new, small and niche farms could use this marketing technique very well.

Tammy Dobbs was very successful using the ezine technique. She lives on five acres in New Lowell, Ontario. She wanted to help local small farms find customers, so she put up a cooperative website at:

http://theruralmarket.com

She began drawing potential customers there by writing short, interesting pieces for an ezine service. "I believe people are genuinely interested in the products they eat these days," says Tammy, "and I think it is just a matter of educating them just a little about fresh products versus store-bought." Tammy wrote a small article comparing a farm-fresh product to that of a typical grocery store, which she says, "took me about an hour to compose, and it has been one of the best things I could have done to promote my website. Most of the guests to my site come through the signature at the end of that article, so it was well

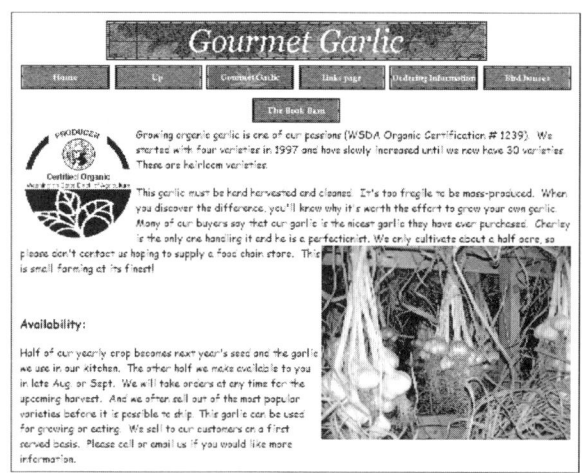

Charley's Farm, www.charleysfarm.com

worth the time it took to put it together." The "signature" Tammy is speaking of is at the end of each article and contains the author's name, farm or other defining information with a link to the author's own website.

Other examples include a California group promoting agritourism that wrote an article about the history and transformation of Kozlowski Farm (www.kozlowskifarms.com), a destination farm where customers can stop off for a picnic and purchase items from the farm store like Raspberry & Roasted Chipotle Sauce, Red Raspberry Vinegar, Red Raspberry Fudge Sauce. From that story, readers are led to a link that takes them to a site offering maps to this and other agritourism destinations in California.

An agritourism guide service in Europe wrote a review of a farmstay in Tuscany, Italy, with a link that led to more information on booking a trip there, and even more agritourism farms in Tuscany.

An eco-destination with its own organic farm on the island of Kadavu in Fiji published an article on how to use coffee grounds in your compost. They commented on how they used coffee grounds this way on their own organic farm and eco-destination, which naturally got readers curious about that eco-destination, and a link from this article takes readers to their website to learn more about it and how to book reservations.

Your articles can include how-tos on planting the seeds, bulbs, or nursery items your farm grows and sells, reviews of good times on the farm and why they help families bond or help children eat more healthily, gift ideas readers can make from your farm products, etc. Another tip for these online "edutising" articles: people seem to have a bottomless appetite for recipes! The main thing is that the articles must contain at least some useful information for the reader on their own, as one might find in a newspaper or magazine. They can't just be a list of your new products or an ad for your new B&B. A look at others' articles will help you pick up the "voice" used to write these. For a good example, see:

www.ezinearticles.com

Though online marketing can go even beyond the basic website, e-newsletter, blog, and ezine articles, these are solid foundations that may need nothing more. In fact, depending on your farm's

Kozlowski Farms, www.kozlowskifarms.com

situation, just the website itself may be all you need, at least at first. One blueberry farmer reported temporarily taking his site down because he received more orders for his products than he could handle. As Jane points out about online marketing and all its many options, if you are a beginner doing too much at first, you can get overwhelmed. "The key is to get started with a website," she says, "and keep it updated."

The website

Your own website must have a registered name that no one else uses. Called the domain, it's the name computer users type to bring your website to their screen. The most logical name would be www.YourFarm'sName.com. "It's best to choose a domain name that is either your farm or business name or one that is descriptive of your business," Jane says. "The actual farm name may already be taken, and that's why a good business description can also be effective: www.bobscornmaze.com (for example)." Heartsong Farm of Groveton, New Hampshire, chose www.herbsandapples.com. Owners Michael and Nancy Phillips combine his history of growing organic apples with her herbal wisdom. Their domain name tells customers directly what the farm specializes in. "Many times these names are selected because of the key words people might use to find a farm or ranch type of business," Jane says.

Domain registration services tell you if your chosen name is available. If not, they often suggest possible alternatives.

Website design has similarities to designing your farm's brochure or roadside stand. Where will the title or sign go? What will the table cover, paper color, or website background color be? But website design also has its particulars. The "home page" of the website—the first page people come to when typing in your domain name—can be considered parallel to a magazine's cover. It should be high in visual appeal and not too wordy or crowded, but with enticing invitations to look deeper. Other web

CHAPTER EIGHT

Heartsong Farm, www.herbsandapples.com

pages that lead from the home page are like pages within those magazines.

"A good website really needs to have offerings much like a magazine," says Michael Phillips. "You want people to flip through the pages, find all sorts of intriguing cul-de-sacs, and then hopefully support the 'advertisers' by ordering some of the farm's products."

Michael's website offers photos of people picking herbs and enjoying the farm, something Jane says is very valuable. "People don't want to look at a building through an open parking lot," she says. "They want to see the people enjoying themselves, picking the fruit, eating the ice cream, etc. Photographs are key to a good farm website." She explains that a major mistake is making the homepage boring. "People using the World Wide Web are used to getting information in a hurry," she says. "They want to get the information, see the photos and make a decision. They are not willing to read a sea of your best prose." Many website owners find that once someone becomes a loyal customer, or has become "hooked" by your website, they then go in deeper for informative and longer text. The homepage, though, needs quicker, visual appeal.

Yet even though homepages shouldn't be wordy, there are verbal essentials. Jane notes another major mistake on homepages is not telling anyone where

you are or how to reach you, leaving out the full address and phone number. "Many farmers just don't realize that people are searching from greater distances than (the farmers') own communities and may not know where they are located," she says. "Maps often times are so local that they don't give a frame of reference to larger highways or north, south, etc."

Since e-mail addresses or even phone numbers are often picked up from the web for unsolicited advertising (SPAM and telemarketing), you might want to look into methods for online visitors to contact you first without revealing this information until you know a genuinely interested person is on the other end. But even this form of first contact should be easy for your readers to find and use, and they should immediately know your farm's country, state or province, and regional location without having to ask.

To prevent your e-mail address from being picked up by the spammers, for example, make sure that your actual e-mail is never posted on your website. Most web design software will allow you to simply put in a link that says, "e-mail us…" which, when clicked on, will open up the viewer's e-mail program with your e-mail address typed in the address field. Spammers can then scoop up your e-mail from this, but this method takes longer and lessens the occurrences. The other option mentioned above is to allow viewers to fill in a form to send you messages and type in their own e-mail addresses first. The message is sent to you without them ever learning what your e-mail address is. You can then decide whether to begin an e-mail dialogue with your actual e-mail address once you receive the message. In both these cases, both yours and your viewers' e-mails are hidden from the web spiders, and most simple web design software will have these features.

For an example of how one farm uses a "first contact form" instead of revealing the actual e-mail

Boistfort Valley Farm, www.boistfortvalleyfarm.com

on a CSA site, visit Boistfort Valley Farm's website at:

www.boistfortvalleyfarm.com

Deeper within your website, you can offer stories of the farm's history, how-to's on preserving the berries you sell, or as on Charley's Farm's website, tips on sustainably growing the products your customers buy from you. Online ordering, if added to your website, will most likely be its own page or pages, with a button on the homepage taking readers there.

"Be creative!" Michael advises. "Your farm and the produce and animals you raise are things to celebrate. Good energy can come across on a well-designed and earth lovin' website. Stay away from the hum-drum, be informative, and make everything you do be a class act."

Just as there are rules for making brochures ready for the printing press, there are rules for making websites ready to be published online, such as files to save and codes that must be typed in. Professional webmasters and designers are available to take your website from its initial design to a published site online.

Barters and trades have also been swapped between web-savvy individuals and the farmer, trading produce for a website. Visit www.eatwell.com to see an example of a website done through barter with a farm customer. If none of your customers are particularly web-savvy, perhaps you can find someone willing to create an easy-to-set-up and maintain website for a reasonable fee. Or, you may be lucky enough to have a web whiz within your own family or farming partnership. One tip—and I'm only half joking—is to look for the nearest 14-year old! If you choose this option, though, be sure to stay on top of your website and at least learn the basics yourself. One farmer who turned his website over to a young son found hidden, war video games the boy had created himself secretly imbedded on the farm's site.

Remember that a basic farm website doesn't always necessitate the need for a techy or professional designer, however, as there are lots of easy-to-use web design software programs that don't require special coding or advanced computer expertise. Do a Google search: "easy to use web design software" or look in computer magazine reviews, or visit the software section of any major office or software retailer to find basic website design software.

With a website ready to publish, a hosting service then publishes your website onto the World Wide Web, making it possible for computer users to type in your website's name, and have it appear on their screen. And just as you can't, in general, back up to any farmers' market you happen to come across and sell from your pickup without first registering and paying a farmers' market fee, your website host will usually charge a monthly fee to keep your site up and available to the world. Free hosting is available, but sometimes at a different type of price, such as collecting your name to sell later to others, or allowing advertisements to pop up on the screens of your website readers. The ads could be anything from dating services to get-rich-quick schemes. Also, you may want to choose a hosting service before the actual web design takes place, as some hosting sites have templates or restrictions that may cause you to need to or want to redo your original. Once published, you or the webmaster should continue to keep the site updated. (Eric Gibson, publisher of this book, recommends fatcow.com. "At $99 a year, it has a lot features, including a shopping cart," Eric says.)

The E-Newsletter

The e-newsletter may contain crop ripening updates, new activities available for your farm's B&B or harvest festival, price changes, and so on. It may also include a farmer's journal (see Chapter 10). You can either just publish it on your website, leaving it to the readers to go there and find it, or you can send it out either via e-mail or an RSS feed. For either, your website should offer new subscribers an easy sign-up.

If choosing to send out your newsletter through e-mail subscribers, it's important to promise (and deliver) an easy option for unsubscribing, and guarantee that the new subscriber's information won't be given out. Even if you're just starting out with your first four subscribers whom you plan to send the newsletter to manually, don't send the newsletter out to them all at once in a manner that lets the subscribers see the others' names and e-mail addresses with a carbon copy (CC) option on your e-mail account. Instead, send them out all with the blind carbon copy (BCC) option. Potential new subscribers can be hesitant to open the door to unwanted solicitations in their inboxes, and they need this reassurance to take a chance on you. All subscribers must be those who choose to be on your list, or your newsletter can be considered SPAM—those dreaded, unwanted inbox sales pitches.

When subscription numbers grow, online services will manage your e-newsletter for you, including mailing out many at once, adding new subscribers, double-checking to make sure it was they who actually signed up in the first place, and handling those who choose to unsubscribe. "I pay $15 per month for my newsletter hosting site," says Bill Gluth of "Develop You Vision" mentioned above. Another option includes purchasing newsletter software for your own computer. "Personally," Bill says, "I am not a big fan of software that manages your newsletter. You still have to manage (the newsletter), monitor it, and that takes time, which can be better spent running your business." Bill also feels that setting up a manual newsletter from your own inbox takes too much time away from the actual farm business, and can be difficult if not impossible once the list grows. A better option may be to start right out with a subscription service.

The voice and content of the newsletter is very important. The newsletter is not supposed to be a thinly disguised advertisement. Readers have ads shouted at them daily, and need something of value for their time, or they'll stop reading. Bill says to emphasize telling authentic stories that express true talent and passion, and to do it with consistency. "So many people under-perform only because they just don't communicate (successfully through newsletters) consistently," he says. Readers of farm newsletters want good, solid, usable information, such as healthy recipes for your organic vegetables. And, as described in Chapter 10, they want your farm's story, a journal of life on the farm.

RSS feeds are another method for syndicating content over the Internet. Subscribers who use RSS readers receive the new information every time the feed is updated. At this writing, more people are familiar with e-mail newsletters, but RSS feeds are rapidly becoming more common and user-friendly. In one way, RSS feeds are less valuable than e-mail newsletters right now in that they aren't as familiar to the mainstream, nor to you for setting them up. But they have advantages as well. A new audience may prefer them because one doesn't have to give away private information to subscribe. There's no SPAM involved, so subscribers don't have to worry about SPAM filters blocking their subscription delivery as sometimes happens with e-mail subscriptions. At this time, it's suggested that both options be offered, and then you can monitor what the response is down the road. For setting up an RSS feed option on your website from scratch, the do-it-yourself RSS feed arena is improving. There are online tutorials for setting them up. Some are loaded with techy jargon, and some are more layperson friendly. If you have weblog (see below) capabilities, these sometimes have automatic RSS feed capabilities.

If you create e-mail newsletters with very rich content, such as with sound, video or large color photographs, you might want to consider the option of just sending out an enticing opening to your newsletter instead of the whole thing, with a link to the online page where it can be found in its entirety (e-newsletters should be no more than 250k in size.) This helps keep huge files from filling up or sometimes clogging your subscribers' inboxes.

How often should you send out a newsletter? Patricia Cobe, co-author with Ellen H. Parlapiano of *Mompreneurs®: A Mother's Practical Step-by-Step Guide to Work-at-Home Success* and *Mompreneurs® Online: Using the Internet to Build Work@Home Success,* warns of overdoing it, especially if using the e-mail option. "I think e-newsletters can be a great marketing tool if they're not overdone. A newsletter should only be done if it offers something valuable and different… and should never appear daily. We all have too much clutter in our e-mailboxes already!" Patricia and others suggest once a week to once a month.

According to *The New Farmers' Market* (www.nwpub.net), "The website of the Crescent City Farmers' Market in New Orleans (www.loyno.edu/ccfm) jumped up to about 1,500 visits a month in 2001, a remarkable number, considering that when it first went up three years previously, it got about 50 hits (visits) a month. Market

> *Blogs are sometimes thought of as online farm reports of customer on-farm activities or ripening times that guests can open, read and even add to with their own comments. Little bits and pieces written two or three times a week by the farmer supplement or replace the e-newsletter—and some say are the hottest thing in online marketing.*

director Richard McCarthy attributes the jump in website traffic to eMARKET, a weekly e-mail version of the market's newsletter.

"When we first set up our website," says McCarthy, "it was a typical nonprofit organization website with a nice logo and a few nice pictures, but very static and not a lot of information. When we started the bulk e-mails, our visibility on the Internet grew remarkably." (A service called listbot.com, at www.listbot.com, can help you set up an e-mail mass mailing).

"About 700 people get the weekly e-mail letter, so it's a highly efficient way to reach our customers on a weekly basis instead of mailing our monthly quarterly newsletter to 6,000 people.

"We haven't been able to tell exactly how many people come to the market from the website," McCarthy says, "but we do know that it includes a lot of tourists."

The weblog or "blog"

The blog is a place where viewers can count on finding something new in bite-sized pieces, and where, if the blog owner chooses, they can add their own comments. Your blog page can be hosted by a blog-hosting service with a link from your website, or it can be part of your own website if your website host offers this service. Like the website, readers can go to the blog's site to read its content. Blogs can also be subscribed to with RSS feeds.

While websites should be updated and progressive rather than static, it can take time to re-design pages. But once a blog page has been set up, it's a matter of typing a few words from any computer, clicking a button, and the blog is now online and available to the world. It's very pleasant to communicate with readers this way. Those who feel blogs are good marketing devises feel they attract more search engines than static or less active websites. Search engines can be considered online Yellow Pages, searching the entire World Wide Web for any topic that's typed into its field. "Blogs are indexed by search engines daily because of new content being posted regularly," Bill says.

When allowing others to add comments to your blog, you offer a way for them to feel connected and heard, and for the farmer to receive marketing feedback. However, be aware of hosting a blog that allows anyone from anywhere to add comments without becoming a registered "member." Remember being nine-year-olds and calling strangers on the phone to ask if their refrigerator is running, then when they replied, "Yes," telling them they should go catch it, and hanging up? This practice now includes teens and grownups online, and the wording is harsher. With blogs, you can eliminate comments, or even disallow all comments if you simply want to use your blog as a progress report or supplement to your e-newsletter.

In searching for a blog hosting service, note that there are paid hosting services and free ones. Check free services first to make sure annoying advertising doesn't appear on your readers' screens, and remember they may be collecting names that they then sell to other web entrepreneurs (including, probably, some unwanted ones). Some free or less expensive services don't allow you to be as original or creative as you were with your website. You may be able to upload your farm's name and logo, but there are often a set group of templates from which to choose.

Once set up, your blog posts will accumulate, giving you options such as showing the latest one at the top of the most recent ten or so, and storing past ones in archived files.

Charley's Farm has expanded each year, offering more products on their website, with Ginny able to manage the site according to her work schedule. While there is something very unique and grounded about farmers, we can remember, like Ginny, to explore new marketing territory as time marches on. When it came to succeeding at online marketing, Bill Gluth did something unusual in order to discover his "edutising" idea: he studied the completely unrelated but booming motorcycle industry. "One thing that everyone misses," he says, "is how you can use examples from other highly successful industries. New ideas come from unexpected information sources." The Web is here for the long haul, but even though methods of marketing with it will change as well, throughout time, adaptability has been one of the farmer's best business friends, something that never needs to change.

More ways to draw traffic to your website and weblog

Drawing traffic to both your website and weblog is similar to the networking that goes on at the watering hole. Find other websites or blog groups who will exchange links with yours. List your farm on sites that lead people to farms by region and product or service (see sidebar, "Online Centers" in this chapter). Also, look for sites that list festival and tourism destinations for free that are related to your farm's agritourism, but not directly agricultural: If your ag-festival invites a bluegrass band, or sets up craft booths to teach children how to make fairy wands from your lavender, do a search for sites that cater to bluegrass fans or Celtic festivals and gatherings.

Work to get a link from the entertainers' sites that are set to perform at your festival. There are people that follow specific bands. You may want them to come to your festival or event.

Don't just do link exchanges indiscriminately, however. Exchange only with quality websites that are relevant to what you offer. The search engines' ranking tools have become increasingly sophisticated, and when it's obvious that a website has a large number of "junk" links, it counts as a big negative against that website. Contact your tourist boards and let them know about your market website. Exchange links with local restaurants on your market website because people who are traveling often will look up restaurants in the city they are traveling to.

Find online communities or chat forums of like-minded groups where you're allowed to put your site or blog link as your signature at the bottom of your posted messages. Patricia Cobe, the Mompreneurs® co-author, says that farming, gardening and home-based food selling businesses can definitely benefit from joining online message boards or communities. "We've found that to be true in our Mompreneurs® Meeting Place Boards that we now run on our website (see "Contacts" sidebar)" she says. "Many of the women who have met there have generously shared leads and referrals to others, and the exposure has gotten participants new clients and customers. We encourage posters to talk about their businesses and include a tagline with each post that includes the name of their business and their website. The family-owned MaryJanesFarm of Moscow, Idaho, offers numerous agritourism services, including a wall tent B&B (see Chapter 14) and an on-farm farming and country living school. The family also operates its own online community, with chat forums, drawing interested readers and customers from across the country. The business also publishes books and magazines, and manufactures and distributes organic, dehydrated camping and convenience foods wholesale and retail.

But even if choosing just to post occasional comments on others' sites to help get yourself

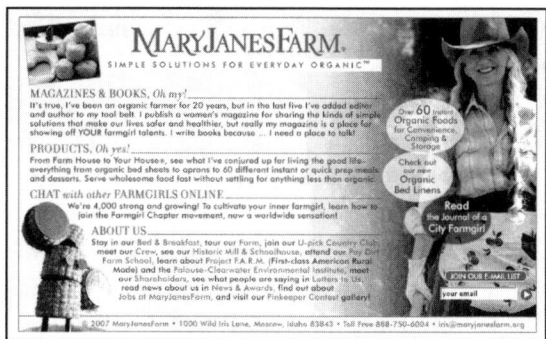

MaryJanesFarm, www.maryjanesfarm.org

known, realize that most people can see through a post that's "commenting" only for the purpose of selling a product. If the chat topic is, "Does anyone have recipes for lentil soup?" try to be a little more tactful than, "Our farm B&B now offers fresh-from-the-oven apple bread from our own antique apples, delivered outside your bedroom door, and we have a discount on reservations during the month of November for the next two weeks, no, I don't have any lentil recipes."

If you choose to have a blog that allows outside comments, watch for this type of promotion happening too often to your own blog site as well. "Today, many people have begun posting on blogs as a way to promote their business," says Bill Gluth, the owner of Develop Your Vision. "This typically takes the original topic off track and does not serve the readers of the blog. All blog publishers need to decide for themselves if comments add to or detract from their focused goal, and use comments accordingly." Understandably, you'll lose readers if they feel they're just going to see ads.

Get in the habit of promoting your website anywhere you'd put your phone number: on your brochures, business cards and advertising, in news releases, on the side of your truck, etc. In addition, whenever you submit an educational article or farm-fresh recipe to your local newspaper, be sure to include your website address in your author's info at the bottom of the article.

There are many more low-cost or even free services that can help get your farm's business known to new customers and remembered by past customers. When someone types "heirloom carrots," into a search engine, and this is your farm's specialty, you'd like your farm's name to come up in the list of options from which the computer user can choose. Search engines, such as Google, will hunt you out for free, and a good webmaster will know how to create website content so your site will be found more easily. While some have tried online "pay-per-click" advertising with little success, it has worked for others. Pay-per-click's appeal is that you pay for a listing that comes up higher on the search engine's list, and pay only when someone actually clicks over to your site from that listing. "We advertise our garlic on Google," says Ginny Hein, of Charley's Farm, "since that is a pay-per-click type of advertising, we can limit the search words to people who are growing (garlic), and that reduces our overall advertising costs. Pay-per-click type advertising has been our best source of advertising for the garlic." She went on to advise, "Limit search words as much as possible to reduce costs." Though the less words used, the more money saved, the Heins chose to use the words "growing garlic" instead of just "garlic." Ginny says that when they just used the word "garlic" they got a lot of clicks (which they must pay for) but few buyers. In their case, their best customers are those who want to actually grow unique garlic varieties themselves and purchase those bulbs from Charley's Farm. ❧

Chapter Resources

Resources

Eckert AgriMarketing
Jane Eckert
8054 Tesdale
St. Louis, MO 63130
314-862-6288 office
314-721-0825 fax
618-593-6129 cell
www.eckertagrimarketing.com
www.farmwebdesign.com

Develop Your Vision
Bill Gluth
11536 W. Bighorn Ct
Surprise, AZ 85374
623-210-3203
www.developyourvision.com
www.billgluth.com (BlOG)

Mompreneurs® Online
Work from Home Strategies
www.mompreneursonline.com

Small Farm Central
Provides website and ecommerce web services to direct-marketing farmers and also has free info for farmers about setting up farm websites, posting farm blogs, etc.
www.smallfarmcentral.com

Online centers where your farm may qualify for a listing

(Note: Some are free of charge.)

www.agritourismworld.com

www.biodynamics.com

www.chefscollaborative.org

www.farmerchefconnection.org

www.farmstop.com

www.foodroutes.org

www.landstewardshipproject.org

www.localharvest.com

www.sustainusa.org

Also, check your local Chamber of Commerce, Tourism Department, Department of Agriculture producer listings, Growers' Associations, shared-interest groups related to your farm's agritourism, general event and festival listings.

Free Promotion 9

A well-written news release can generate a rippling form of free promotion. It took only one article in our local paper about a classroom field trip to our farm for word-of-mouth marketing to jump-start. It ignited inquiries from people wanting to know about our other farm services and products.

News periodicals, TV and radio stations are always looking for the latest story to keep their audience interested, and they use news releases as a major source for their on-going story search. The news release serves to invite them to send a reporter to cover the story in full. It isn't the entire news story per se—the staff of the form of media you've attracted will write the actual story. You or your staff will create the news releases with hopes of attracting them to your farm's agritourism activities. Make them sound newsworthy rather than hyped PR pieces: "Make it news, not advertising." Journalists and editors are always asking themselves: "What's in it for my readers/viewers?" and so are usually quick to spot the amateur news releases written by those who simply want free promotion. On the other hand, they're fine with the fact that a story on you will be of promotional service to you, as long as it's of genuine interest to their audience.

Collecting addresses for news releases

You'll most likely want to send out news releases to as many media outlets as possible. Your local library will usually have directories for media outlets. Look

> *Make your newsreleases sound newsworthy rather than hyped PR pieces: "Make it news, not advertising." Journalists and editors are always asking themselves: "What's in it for my readers/viewers?"*

for local and regional newspapers, business journals, club and church newsletters, radio stations, TV stations and online sites that report local news.

Collect addresses and find the specific department you want to contact for those large enough to have several departments (such as the Food department of a large newspaper if your news release is announcing your farm's latest gourmet vegetables). Also, it's helpful to address the news release to the actual name of the reporter or editor you think should receive the news release. So write these names down as well while you're collecting contact information for media outlets. Otherwise, your release can be filed away and bypassed.

If you want to go national with your news releases, there are a growing number of services online that claim to send your news release out to many national media outlets for a fee. Some of these have proven fruitful; others have been a waste of money, if not a downright scam. If such services promise large numbers of your news releases will be sent out

Front cover article in Acres U.S.A. Magazine, in which this book's author, Barbara Adams Berst, writes about Home Sweet Farm. This is the kind of free publicity that is invaluable! www.acresusa.com

via e-mail, check to find out if the types of media are those that would be interested in your story. Any "service" can collect the e-mail addresses of every tabloid, fashion magazine and of course, Beer Lovers and Swingers of America, and send out the news release on your Vermont bamboo farm's open house to "large numbers." Check to see if they have statistics on how many of their news releases are followed up on and if they follow through to see if editors

appreciate and accept these electronic news releases, or if editors filter or delete them automatically as SPAM.

If the news release service promises, instead, to put your news release up on their website and claims their website is perused regularly by the media, check to see what media they're talking about—the definition of "media" is changing, now that just about anyone can become an online publisher, so see if they can supply actual data on hit numbers by the legitimate media to their website. Also, consider checking if and how often individual news releases published by their company are actually found among the crowd of other news releases. Don't fall for excuses such as, "That would be impossible to for us to check." Of course it's not impossible. The marketing industry has tracked results of various campaigns for years and has gotten it down to a science.

The Internet has the potential to greatly improve the way our media finds its stories, but scammers will take advantage of any opportunity, including the Internet. Don't ever go by any mass-distribution service's data as supplied by them alone—find if and how such statistics can be verified by a third-party independently. Ask any service you are considering using for referrals of satisfied clients, and also ask them if their services have been reviewed or rated by a third-party. Also ask them for the names of three or four media outlets they send releases to; then call those media to find out if they do, indeed, welcome or use news releases sent by the service you are considering using. Better yet, "work backwards:" Call half a dozen or so of the media you would like to receive your news release, and ask them how they typically receive news release they end up using—from news release services? Or sent to them directly by individuals? You may want to start out collecting national media addresses and editors' names on your own, and send print news releases directly to them until you find something reputable worth using.

Making the news release professional: Technology and media evolve and change almost daily, and as one might suspect, there are differences of opinion on what makes a news release professional. Most local and regional services accept privately created print news releases. Some encourage e-mail news releases; others heatedly discourage them. See Sidebar for a basic template for a printed news release that most agree will pass the test and could be appropriately adapted to an electronic news release.

The headline

The headline should grab the reporter or editor's attention. To do this, write a headline that the audience of his/her periodical or station would find compelling. The following categories with examples may help you tap into the headline potential of your own farm:

- *A benefit for the audience:* BRING THE KIDS FOR MELT-IN-YOUR MOUTH HEIRLOOM STRAWBERRIES AND PONY RIDES
- *Curiosity:* IT ISN'T OFTEN THAT YOUR BREAKFAST COMES RUNNING UP TO YOU, SMILING (story about a local goat dairy, and how the goats expression appears to be a smile)
- *A question:* DO YOU REMEMBER WHAT REAL FOOD TASTES LIKE?
- *News breaking information:* ORGANIC PRODUCE SHOWN HIGHER IN NUTRITION, UNIVERSITY STUDY REVEALS
- *A testimonial:* "MY CHILD NOW LOVES SPINACH!" (story on a U-pick vegetable farm where children harvested and ate fresh spinach right from the market garden)

Always put yourself in the mind of the reporter, and of the audience. If you can fill their needs to draw an audience, they can in turn fill your needs for promotion.

More headline ideas include:

- RARE BREED FARM ANIMALS RETURN TO THE AMERICAN FARM
- MICRO-FARMER FEEDING THE NEIGHBORHOOD
- OLD MCDONALD'S FARM HAS RETURNED
- RARE FAMILY HEIRLOOMS GROWN BY LOCAL FARMER
- URBAN FARMER IS UPSCALE RESTAURANT'S BEST-KEPT SECRET

Nate O'Neil, owner of Frog Song Farm, describes his farming operation to journalists and photographers.

Template for Printed News Release

NEWS RELEASE

FOR IMMEDIATE RELEASE: (these three words tell them the story is available now, rather than something seasonal they should file for later.)

CONTACT: (Actually type the word "CONTACT" in all caps, then follow it with your full name in appropriate upper and lower case)

Your farm's name

Phone number and/or voice mail number

FAX Number

E-mail

Website URL if applicable

Headline (see below for more details)

The Lead: (City, State, Date)—First paragraph that tells who, where, what, why and when. The "Grabber" is a short attention-getter, supplementing the Headline, that will "hook" the reader immediately and draw them into your story.

Quotes and Credentials: An important quote from the farmer or perhaps a guest at one of the farmer's events, and the title/credentials and quickly summarized human interest story of the farmer.

Body of the news release.

Call to Action: Further details on what readers should know and do to participate.

END—(Actually type the word "END" at the end of the news release in all caps. This tells them there is not another missing or lost page somewhere.)

You don't have to be doing something rare or spectacular for human-interest articles. You can also attract attention by touching human hearts:

- EX-FARMER RETURNS TO HIS ROOTS
- WHEN BIRDSONG AND SUNRISE ARE YOUR "BOSS"

If you're aiming for a national audience or even international, note what the specific media you plan to send the news release to likes to report on, and choose a theme (or 'angle') based on that.

- SMALL ORGANIC VINEYARD TUCKED AWAY IN THE HILLS OF WALLA WALLA WASHINGTON ATTRACTS WORLDWIDE VISITORS (Travel magazine, readers want unique places to visit)
- SINGLE MOTHER FINDS HOME BUSINESS NICHE TEACHING ITALIAN COOKING ON HER ONE-HALF ACRE HEIRLOOM TOMATO FARM (Woman's magazine, women want inspiring examples of how other women make their lives successful)

Fictional News Release

NEWS RELEASE
FOR IMMEDIATE RELEASE
CONTACT: Judy C. Abrahms
Lavender Lane Farm
Phone: 089-123-4567
Fax: 089-765-4321
Judy@LavenderLaneFarm.com
www.LavenderLaneFarm.com

HISTORIC DISTILLER ON LAVENDER FARM BROUGHT TO LIFE AGAIN

(Lanesville, OR, July 15, 2010) An historical distillery demonstration will be held at Lavender Lane Farm using a refurbished distiller found ten years ago hidden in a shed among the fields of the farm. Owner Judy Abrahms steps into the past to distill her farm's lavender essences which, for the past two years, she has shipped out to select cosmetic companies around the world. And on July 25th, she invites the community to experience history with her.

"This was the most fun our family had all summer," said a visitor to last year's demonstration when Abrahms, a third-generation lavender farmer, put on the lavender distillation demonstration for a smaller crowd of her own church members.

It was then that Abrahms realized many people don't know where their food, or their scents, come from. She vowed to make her blooming lavender fields and distillation demonstration open to more people in the years to come. It is very similar to the process her grandfather used when he grew and processed lavender on the very same farm.

Lavender Lane Farm will open its gates to the public from 10 a.m. to 6 p.m., when the fields are expected to be in full bloom. There is a $2 entrance fee for age 12 and over. Children younger than 12 may enter free. All children should be accompanied by an adult. People may pay at the farm, but can also purchase tickets ahead of time at the Lane Bookstore and Lane's Flower Mart. For driving directions, visit www.LavenderLaneFarm.com.

♦ THE IMPOVERISHED LEARN TO FEED THEMSELVES ON NON-PROFIT ORGANIC DEMONSTRATION FARM (International magazine, readers want stories on how the world is improving with methods that can be duplicated in many locations around the world).

Follow-up

Once news releases are sent out, most journalists I've spoken to agree we don't like pushy follow-ups. As both a farmer and journalist, I've dealt many times with public relations departments contacting me on their client's behalf to try to sway me to say what they want me to say, and I have to refuse. Journalists are under oath to serve the public, not the person they're writing about. That's why their story about you is so valuable to you when it's positive. It gives you more credibility than a paid advertisement. Readers, viewers and listeners are supposed to be given the real story, not the paid promotional story. In fact, no journalist should ever tell you he or she will write a favorable article about you if you pay them with either cash, services or goods. Journalists are paid by the media they write and report for, and are never supposed to accept additional payment or goods for writing favorable reports. Certainly, they can be allowed into your harvest festival for free as a courtesy. Or, let them come visit your farm and pick a vine-ripened heirloom so they can report first-hand on what you have to offer. But accepting payment from a citizen to write a favorable report is considered fraud. Do keep in mind that they can also write what they see as negative, so show your farm at its best when they come to call, just as you would for your customers.

If you are contacted by a journalist for an interview, congratulations! Usually, they'll ask you many questions, and just weave a few of your answers into their story. Be careful of "padding" your answers with promotion before answering their questions. I sometimes interview farmers with questions such as, "How did you first decide to raise rare breeds?" and receive answers such as, "Our farm is the best farm to buy this rare breed from, our prices are the best, our quality is the best, no one should purchase them anywhere else, and we decided to raise them after seeing them at a fair." I then ask more questions to dig out a deeper answer to my original question, "Which fair? What did you learn about the animal at the fair that drew your interest?" Then I drop the "padding" for the actual quotes in the article.

Also, don't try to dictate the article to the journalist, or suggest how they do their job. As an example,

Is there such a thing as too much media?

Some say they would die for this "problem" but it can happen. Hopefully, most journalists are well-intentioned, but some are less so. From the very start, do stay in charge as to which reporters you spend time with. But especially if too many requests come in, use the following to determine how to handle them:

1. Does the media reach the crowd you're seeking? Who is their audience?

2. Does the media, and hopefully the specific journalist, have a reputation for writing fair, deep profiles of the people they interview, rather than superficial travel pieces, sensationalism, or sarcasm (ask them to send clips of similar articles they've written, or point you to websites that contain them).

3. What is the angle, or "theme" of the story? Does the journalist want to show the real story of local farming, or the history of your crop, or the experience of your farmstay? Or, does he or she want to write a piece about your entire county, asking you to do the work for him/her on providing facts about the county's weather, its history and other destinations?

5. Is this disguised journalism? The field of journalism is, in general, for profit, just like farming.

But is a profit-making tourist company, unrelated to the legitimate, unbiased media, trying to draw people to their resort by showing them the quaint farms in the area in the name of a fabricated news story or documentary? Or, are they using you to cash in on a trendy "local tradition" image to their advantage (gaining more high paying tourists to their resorts), and not yours? Will any of these high paying tourists ultimately benefit your farm in any way? Will they possibly pay to come to your on-farm cooking classes, for example?

I tried to interview a tourist expert in France about their country's agritourism, telling him that I write about agritourism worldwide. I was quickly told that I should only write about France's agritourism, no other countries, because the French have the best agritourism in the world. I dropped that particular interviewee completely, but certainly didn't hold it against French people in general, most of whom I found extremely gracious and helpful.

Journalists often can see beauty and interest on their own that they'll interject into your story if allowed to look around and see it for themselves without having to fight off an over-zealous interviewee. On the other hand, don't be afraid to politely ask if you can tell something of interest you feel they're overlooking. "Did you see this? This is where we compost and where visiting school children learn about worms."

If there is not an immediate response from your news release, wait a few weeks; slightly rewrite and update the news release; see if there's another reporter or editor to direct it to, and send it out again, without any complaints about not hearing from them earlier. Media personnel get busy, get fired, get sick, and get transferred to new locations just like the rest of us. Also, your story may have reached the wrong editor, but has been passed on to a different one. Our farm was contacted by the local business editor, whom we had never contacted ourselves, after a news release was sent to a different department. Remember, also, that if there's a lot of competition in your location, the old advertising code may also apply to news releases. The person you're trying to attract may need to see the name seven times before taking action. ❧

The Farm Journal 10

Telling the farm's story through a farm journal can be a remarkable method for gaining and keeping farm customers. Heidi & Mike Peroni, owners of Boistfort Valley Farm, a CSA in Curtis, Washington State, report six things that bring their customers back to their farm. One of them is described as "Connection with the producer (aided greatly by receiving the Mike/Heidi musings about what's going on at the farm)." These "musings" are in the form of little written notes and thoughts added into their newsletters. Customers repeatedly remark about how fun it is to hear about the work that goes into the produce they receive.

Note: A word about the difference between a farm journal (also called a "blog") and a farm newsletter: The farm newsletter is often more about facts your customers need to know to purchase your farm's products or services. It tells your customers when new crops have ripened, the price of next year's CSA membership, the dates and fees of a farm event, and sometimes recipes. The farm journal, in contrast, is more of a diary of your own thoughts and perspectives of every day farming, such as how happy you were to see that the new fence seems to be keeping deer away from the ripening melons. The newsletter and farm journal can certainly be combined, and sometimes overlap.

A recent article in the Financial Times stated, "a host of blue chip companies, including the BBC, Emap and Lego, have discovered that storytelling can help everything from marketing and product development to staff management and branding." Agritourism farms are joining other businesses in this important discovery of the power and value of story, as they aim to connect to the general public on the experiential and emotional level.

The Leopold Center for Sustainable Agriculture has further pointed out how important the farm's story is becoming to new emerging direct markets for farmers, including chefs (see Chapter 15).

For you, the farmer, journal entries can be a natural and pleasant way to present your farm as a one-of-a-kind destination to maintain return customers, attract new customers, and even bring pride to the farming family and others involved in the business, including any farm employees. Your farm's unique story can even help protect you from fabricated farm imitations and commercial competition.

Bill Cummins, President of the Skagit Valley Farm Tour (see Chapter 11) points out that while farm tours and expos are very valuable, there's a deeper side to them that needs to be told. It's a story many are yearning for. "It is important to note that farming is far more than what visitors see on a single Farm Tour weekend," he says. "It is a commitment that lasts 365 days a year through all types of weather and all times of the day and night."

Thankfully, in spite of our increasingly visual society, there are still large numbers of people who know how, and like, to read. This isn't to say that journals can't also be filled with sketches and today, even audio narrations and digital photos. But many

Our corn did prove well, and, God be praised, we had a good increase of Indian corn, and our barley indifferent good, but our peas not worth the gathering, for we feared they were too late sown. They came up very well, and blossomed, but the sun parched them in the blossom.

Although it be not always so plentiful as it was at this time with us… yet by the goodness of God, we are so far from want that we often wish you partakers of our plenty.

– Edward Winslow, December 11, 1621, in A Journal of the Pilgrims at Plymouth (Mourt's Relation: A Relation or Journal of the English Plantation settled at Plymouth in New England, by certain English adventurers both merchants and others.) Dwight Heath, ed. New York: Corinth Books, 1963, p. 82 (A farm journal entry possibly from the first Thanksgiving)

Farm journal notes can be transferred to online farm pages to help current and potential farm customers understand the deeper story of your farm, and this "deeper story" is the attraction and essence of true eco-tourism.

still like to read or hear stories and bits and pieces of real life situations. Your promotional brochures, posters, and your website's opening pages should be heavy with visuals. But deeper down, whether in print form, online, or downloaded onto an ipod, if you enjoy putting thoughts down in writing, let your customers rediscover a farm journal.

Whole New Mind author Daniel Pink states that as humankind moves from a linear era to a more "conceptual" age, storytelling presents the truth of your product, service or brand in an extremely powerful way.

For many farmers, the farm journal has served as one of the most historic, and natural, methods for conveying the farm's story. Throughout history, farmers in countries where citizens were lucky enough to have been taught to read and write, recorded their observations of daily farm life, inspiring insight and conveying new discoveries of working and selling from the land. One Roman farmer recorded his despair that his farm was taken over by the Roman government and was now operated by slaves. A Plymouth pilgrim wrote of the local farms' bounty that year, with which his people shared in a great harvest festival with the local natives one autumn. Farmers who kept journals were in the company of explorers, philosophers, writers, artists, naturalists and sea captains who also wrote down daily activities, thoughts, and observations in journals. Some of history's most cherished personal journals available today include those from Mahatma Gandhi, Beatrix Potter (author and illustrator of Peter Rabbit books), and George Washington.

Most journals still surviving from America's colonization were written by men because boys then were more often taught reading and writing. But some women's journals from that era provide intriguing insight into the daily lives of people. Eventually, a few European journals were published in the early 19th century, which popularized journal keeping on both sides of the Atlantic for men and women during the Victorian era. Then, as the Industrial Revolution expanded, personal journals kept by men described city life and politics, and women's

CHAPTER TEN

journals reflected on domestic life at home. But farm journals from those who stayed on the farm are now priceless tools providing ideas on historical seed-saving methods, crop rotation and the insight that comes from everyday people who wrote down their life experiences. While these often become valuable heirlooms kept within families, they also provide a view of the common human struggles and triumphs of living close to the land for anyone, family or not, who later reads them.

Journals also can be group efforts of the entire farming family. According to Scott County educator Dave Resch of the University of Minnesota Extension Service, the journal of a farm family will almost certainly become more and more of a treasure as the years go by. And, he points out, it's never too late to start your own farm journal. Journals can be simple chronicles of daily life, reflective observations, "inward journeys," or larger, fact-filled farm or garden documents.

Writing journals helps us connect to nature, the farm, and something peaceful within. Your farm visitors then can absorb this for themselves as they read your words.

Farm journals kept over the years can become family treasures, and can be shared in part or in whole in a variety of ways with farm customers.

Keeping a farm journal can be mutually beneficial to the journal writer's own self and farming success beyond its attraction of agritourism customers. It is said that in the writing itself, the knowing will come. When we write more reflective journals, perhaps starting with a prompt such as "today's weather" but not really knowing where the writing will take us, it allows the intuitive part of our brains to connect with the logical, and new understandings, amusing situations, and solutions can appear during the act of writing. Even the type of farm journal that simply states the day's facts, or keeps records of crops and seeds, allows the human mind to expand in valuable ways. School marms from the one-room schoolhouse days say that when children taught younger kids, they then understood and retained the information better themselves. Making knowledge that's inside of us come outside of us for others to learn from (such as in teaching it, or writing it out) seems to be the trick to this deeper knowledge and retention. When we write out facts that we already know in our minds, the information becomes deeper, clearer, and oddly, more expanded. When we describe our experiences in nature, those experiences become larger and our perception expands so that we become aware of even greater circles of nature and situations that might just skim by another person. We become more grounded in who we are and begin to seek and observe even more beauty around us.

Writing out daily activities can also help us gain gratitude for living, which is considered by many to be a very healthy state of mind. We start to see that our lives are an ongoing, unfolding novel. When we re-read of the insurmountable problem described in a journal five years ago that no longer concerns us, we can gain a confidence about future problems. You discover you always owned something abundant, and as they say, treasure attracts even more treasure. And there is great satisfaction in telling this unfolding story to others, having it be heard, and to then

> *The skunks and I have a very comfortable routine worked out in the area where the cats are fed. Young male skunk (so called because he has 'attitude') has finally got the word from the old female skunk not to get excited when the nice lady comes out to feed the cats.*
>
> – Journal entry of Carey Family Farm, Oregon

> *I am enthusiastic about getting the winter greens planted. I also plan to start everbearing strawberries. We have some visiting deer who hope to grab some fruit before Big Dozer, my Retriever/Rott cross, spots them. He chases them from the orchard area. As soon as they hit the driveway or the outside fence he sits and watches to be sure they leave. Several deep barks speed them on their way.*
>
> – Journal entry of Carey Family Farm, Oregon

allow them to support your lifestyle by bringing you their business.

What about venting… gossip… and the like?

You may need two journals. As a facilitator for journal-writing, I always tell the students that deep insight, factual knowledge, great wisdom, and "the dark side" are all begging to be brought out of hiding either for obvious good use, or for release and healing. But the part of the story meant for the public needs to be separate from serious harshness, which might best be written out first in a different way where you have no concerns of others reading it. For example, if you're quite certain that today, a family member or recent farm customer should be locked up in a dungeon for their behavior, it can be helpful to vent this out freely on a loose sheet of paper, and then use that sheet as fuel for the fireplace. Or just keep a private journal for venting. After that, go to the more-public farm journal. Your entry for the day can still be honest, but you may be surprised how much grace, confidence and maturity it reflects once the venting has been released: "Joe and I didn't see eye to eye today, but this too shall pass. I was once 14 like him, and today I remembered getting so mad at my mom when she said I'd had enough oatmeal raisin cookie dough. Enough? I'd just begun! 30 years later, I'm now grateful for being taught how to delay gratification." Family members and others shouldn't be subject to publicly-written ridicule or lashings for the future to read.

And if the journal is a family or group one, Dave Resch suggests that you establish basic rules regarding language use and comments. Inappropriate language and hurtful or mean comments should not be allowed, but each person's entry can certainly be individual.

For journals you plan to share with others, keep entries short and on target. A long, gray, single-spaced block of text scares readers away, while short, quick-to-grasp entries are attractive. If you had really meant to write only a couple paragraphs about the beauty of the first butterfly in the spring field, and it somehow led to ten pages concerning a better way to handle gun control in our country if the morons in charge would resign immediately and hand weapon control over to you, remove the page(s) from the public journal to the venting journal.

Also, journals meant to be made public can constitute a form of official publishing and you can be considered accountable just like any published author. Copyright them to you in the year they are written to keep them from being copied and used by others without your permission, and place a copyright mark on them as you would with your website or newsletter. Also, be aware of making personal threats against private citizens such as your neighbor (even if he did paint his house the color of two-week

DE ARBOL CHILES AT THE MARKET

Joan (and Dan) are bringing a new batch of De Arbol chiles to the market tomorrow in San Francisco. These are the short, hot pain-inducing delights I'd mentioned earlier. You can cut them up for heat anywhere but I like to toast them briefly and then toss them into a blender. Pan roast some tomatillos, onion slices and garlic and then add it to the blender as well. Blend until smooth and you have a delicious Jalisco-style hot sauce.

These were grown and dried right here in Northern California. I'll add them to the website, along with the starch corn and CAL, next week.

03:28 PM in Ingredients | Permalink | Comments (3) | TrackBack (0)

A sample blog from Rancho Gordo farm. This popular blog receives 300-500 unique visitors a day. www.ranchogordo.com.

old spaghetti leftovers); of libel (making damaging statements about others); of plagiarism (stealing other copyrighted work without permission and/or paying its owner for its use), and the trouble you can get into if mentioning involvement in illegal actions or making threatening remarks about your government, even if in jest (notice the absence of any examples given here).

Setting up a personal journaling routine

Journal writers use various methods to keep up with their entries. Some choose a daily time, such as at morning tea before others arise, in bed at night, or during a noon meal if you have enough solitude at that time. Field journals go right outside with you, and writing entries may become a routine right after gathering up tools and taking off the work gloves.

Some prefer to just chronicle critical events or to write only when motivation strikes. But it's easy to forget to spend time with the journal without a routine. If you prefer to operate without a set time for writing, consider laying your journal out with a fresh pen or pencil where you often sit to relax or refresh yourself with food and drink. The urge to pick up the journal and write may be irresistible at these moments.

Starting points and prompts

During a journaling workshop, one highly intelligent business man was asked to begin writing anything he wanted on his blank page. He broke out into a sweat, and couldn't put a single word down on paper until finally, he scratched out… "Help!"

When this happens, it usually means the "logical editor" rather than the "creative intuitive" part of the brain is trying to take over the creative process. Let the logical editor help you with spelling and grammar, but it will block your deeper insight and motivation to write if allowed to take the lead in creativity. Instead of inspiring you, it will remind you that if you put a single word down on paper that's misspelled, or if you don't maintain proper margins, Mrs. Beastly will mark your paper with a red F and send it home to Mom and Dad. Writing blocks also can mean the opposite, overwhelm. There are so many things you want to write, you don't know where to start, so it's easier to just not start at all.

To get past either of these—logical editor takeover or overwhelm—at least at first, start each writing session with a pre-designated theme or prompt. A summer theme could be, "The first new blossom I saw today was…" or "The first sound I heard on the farm this morning was…" Other seasoned journal writers love to collect and use surprise "prompts" which they then draw from before each writing session. These can be printed words or phrases, or even images cut up and put into a box or envelope that you draw from at the beginning of each writing session. If you draw the phrase, "How would a passerby describe my farm today?"—where will that prompt eventually take you? You will most likely be very surprised at what wanted to emerge, with that phrase simply giving you leverage to begin. That prompt is capable of leading every writer down a

Blog posted on the Tiny Farm Blog, tinyfarmblog.com

different road. One may end up remembering passing by a farm in childhood, and the original spark that inspired him or her to eventually own a farm. Another may recall he was going to alert farm customers to the opening of the first cut-flower blossoms.

As a starting point for a new farm family journal, Dave Resch says to encourage everyone in the family to get involved by asking all generations of the farm to share a few of their special memories about the family and the farm. Once started, he says that all family members can then continue to contribute special thoughts and events they would like to share with the outside world and future generations.

Sometimes, journal writers will benefit from prompts or designated starting points in odd ways. I've seen writers experience complete inertia when trying to figure out what to write about. Absolutely nothing will come out of their pens. But when given a prompt or "assignment," it is as though a stubborn inner voice is so opposed to being told what must be written about, that a flood of previously non-existent writing ideas pours forth in an effort to protect them from the assignment and being told what to do. When this happens, enjoy!

Sharing the journal with farm customers

When customers will be involved in extended farm stays, such as with farm B&Bs, it may be appropriate to offer a sitting area where past and current farm journals are available. Your journal can be bound in numerous ways. A small, spiral notebook can open flat for easier outdoor writing, and travel with you along with other farm and gardening tools. Those who primarily write indoors, or want to bind them in attractive ways for their readers, may choose to keep their journals loosely on unbound pages, and

then bind them as a New Year's Day tradition. Artistic binding ideas for loose pages include using decorated parchment card stock as a cover tied at the hole punches with raffia. Or, some libraries and other craft class outlets offer courses on exotic, hand-binding techniques, and even traditional cloth and leather hardback binding of single books. Many copy services provide lamination for covers and plastic or wire-spiral binding.

For those who prefer starting out with already-bound, blank journals, hardbound, blank books are sometimes sold at bookstores and other gift and specialty shops. Many are beautiful, with cloth, leather or jewel-look covers.

If you give out printed newsletters, these can contain the latest entries in your journal for that month. "February's Musings" can be sent along with monthly newsletter items such as farm activities available to your customers that month and harvest dates they can expect.

I am happy to be part of a system that helps get fresh food to families, and this year more than ever I feel a sense of being connected to our customers. The CSA concept is often traced back to a group of women in Japan who realized a need to bring farmers and the communities in which they lived together. Loosely translated their idea meant "putting a face on food." Not only does our CSA allow us to deliver a great diversity of vegetables to you fresh, it also brings us together. If I have done my job well you should always feel like you know what is happening on the farm, the farm you are directly supporting.

The local co-ops are getting on the bandwagon here, too. We supply produce to both the East and Westside stores in Olympia, and we were recently asked for some farm photos, which will be displayed on their new digital scanners. I really believe we will all be healthier for it. You must feel a sense of warmth and confidence knowing where and how your food is grown.

We also feel better and more fulfilled knowing who we are feeding. We sometimes chat at night: "Oh I hope Art doesn't think these are too bizarre," "I wonder why Mary Lou doesn't like cilantro," "I bet Mack will love these." Putting a face on food. I hope my face goes on a slender head of romaine and not a carving pumpkin.

– Journal entry of Boistfort Valley Farm, Curtis, WA

www.boistfortvalleyfarm.com

*A*ugust is a beautiful time of year. The summer vegetables are in full swing, the weather is sunny and warm, and the fall crops are growing in leaps and bounds. At the same time, hints of fall are sneaking into our days: in the mornings, the tiniest chill is in the air, and a certain crisp smell that signifies the changing of the season. We have also been caught outside after dark a time or two, unaccustomed to the hasty arrival of evening. It is an exciting and sorrowful feeling simultaneously. While we enjoy fall and the slow down that accompanies it (less weed pulling!), we start to think about the inevitable W-word (yep, Winter).

– Journal entry of Boistfort Valley Farm

*T*his morning, when Mike went out to pack boxes, he came back in almost immediately. I had just settled myself in at the computer when he leaned in the doorway and said, "You'll never believe what's in the field. It's giant!"

I corralled the dogs and ran outside to find a huge bald eagle standing in our south field. In my paranoia, I hoped he wasn't eating the onions (it was early, and I hadn't had any coffee yet). The eagle tolerated my presence for a minute or so, then retreated to a nearby tree. I was awed. From a hundred yards, the bird looked huge. We often see them flying, but rarely catch a glimpse of them on the ground.

– Journal entry of Boistfort Valley Farm

Also look for ways to incorporate your journal keeping with your other avenues of communicating with customers. For example, you might have a small space in your newsletter (including an e-newsletter if you have one, as do Heidi & Mike Peroni) that contains highlights of recent journal entries (with your website address where readers can find the full text). Or if you sell at a farmers' market in addition to your agritourism, and have a bulletin board set up in your display area, photos or short quotes from your journaling can be placed on your bulletin board—this all helps in getting your personal story across to your friends/customers.

Sharing your journal via your computer

A journal can be kept online using one of the many web log services, some of which allow digital color photos to be added. They can be public for marketing purposes towards future customers to draw online attention to the farm. A web log journal also can be used mainly for keeping in touch with current farm customers, such as community supported agriculture (CSA) farm members. CSA farmers know their members consider themselves part of their extended farm family, and access to "their" farm's ongoing journal can be an added bonus to their membership. Members are given a password that allows them to access the online journal. Some web log hosting services also allow RSS feed subscribers (see Chapter 8)

As with print newsletters, your farm journal entries can be sent out with your e-newsletter to subscribers electronically via e-mail, RSS feed, or both. Or, it can just be posted regularly on your website, with archives of past journals available for access. If you periodically offer recipes online for your farmed products to customers, eventually collect them into a farm cookbook sprinkled here and there with quotes from your farm journal, and sell it both online and in print form. A farm's own cookbook with thoughtful musings is a great additional product for an agritourism farm.

> *It appears that my thought of writing a journal through the season was not as successful as I had hoped. Once the bean harvest started, it seemed that we were doing little else. We managed to harvest over 500 pounds of beans in the past five weeks. When the frost came last Saturday night, we rejoiced that the beans were done! This is a labor intensive crop, but worth the effort for the market price. And, who doesn't like green beans? We grow Italian, French Haricot Verts, Golden Haricot Verts and the standard American. The term "beans coming out your ears" is an odd expression, but we have all come to relate clearly to the saying.*
>
> – Greentree Naturals, Sandpoint, ID
> www.greentreenaturals.com

As time passes, your journal will no doubt come to be a journey of sorts that allows you and its readers to know yourself, your farm, and even more about themselves and human nature in a way nothing else can. It may also become a gift of history to your family or other future farmers. To keep such a journal, you simply have no choice but to slow down, reflect, and organize your mind. And this goes hand in hand with why many of your agritourist visitors want to rediscover the farm life in the first place. ❧

Chapter Resource

"Starting your farm blog" at:
  www.smallfarmcentral.com

Networking with Larger Organizations 11

When a local Slow Food convivium (See feature later in this chapter) connected with Edith Walden's Willowrose Bay Quince Orchard, they asked for, and were granted, a tour of Edith's rare quince fruit trees otherwise hidden on her quiet Pacific Northwest island farm. From then on, those who toured the farm quickly spotted her products in gourmet stores and food co-ops, whereas before they had gone less noticed. Edith didn't have to advertise her farm tour, solicit nor organize the visitors—that was all done for her by the Slow Food group. In the chapter, we'll explore the idea of adding your farm to a successful team: networking with established and larger entities, and/or linking with other related businesses.

On a larger scale, every October a group called Skagit Valley Farm Tours, part of the Skagit Valley Farm Bureau, organizes farm tours involving around 15 farms in the beautiful Skagit Valley of Washington State. Their mission statement reads, "To promote Skagit County agriculture by cultivating the bond between farmers and the community through education and hospitality." The Skagit Valley holds some of the most fertile ground on earth, supplying the world with beet and spinach seed, tulips, daffodils, raspberries, red, yellow and white potatoes, broccoli, and numerous niche and micro farm enterprises such as a raw milk goat dairy and alpaca wool farms. It also hosts one of North America's highest raptor concentrations, half a million ducks, 50,000

snow geese, 3,000 trumpeter swans, bald eagles, peregrine falcons, many migrating songbirds, and all five wild Pacific salmon species. A drive through the valley offers views of dairy cows out on green pastures and an ever-changing patchwork of farm fields. Yet urban sprawl is threatening this area, so a group called Skagitonians to Preserve Farmland was formed, and they, along with other farmers and the state's agriculture university, support and form the board of Skagit Valley Farm Tours. The group and festival are financially supported by numerous outside donations. Attendees receive maps for a self-guided tour of each participating farm which describes what's being offered at that particular farm. It may be a chance to feed a goat kid or choose a pumpkin.

Bill Cummins is president of the Tour and owner with his wife, Wendy, of Friday Creek Farm in Burlington, Washington, where they raise traditional colored Boer goats and llamas. He says that the farms involved in this tour are seeing financial benefit from their participation. "One of the challenges farms of all sizes face is efficiently marketing to a targeted audience," Bill says. "The Farm Tour helps to distribute valuable information about the participating farms to people who want to 'buy local.' By working together, participating farms are also able to market on a larger scale that may be cost-prohibitive if done on an individual basis." Bill also notes that the Tour increases public support for farming in general. "The Farm Tour has consistently increased the community's awareness of the need to preserve farmland in the Skagit Valley. Every year we have seen attendance increase, with last year's attendance bringing in approximately 16,000 visitors. By opening up our farms to the community, visitors are able to see first-hand how farming on any scale benefits our county and state."

Examples of other established entities your agritourism farm may want to network with include your local Chamber of Commerce, numerous websites that list festivals and agritourism farms (see this chapter's Resources), B&B associations, your own convention and visitors' bureau, a local garden club that sponsors home and garden tours, your own region's department of agriculture which may publish and distribute farm directories, regional sustainable farming advocacy groups, and regional groups promoting the unique, regional flavor of its own independently owned enterprises, such as artisans, local restaurants, and of course, agritourism. An example of the latter is Tumbleweed Tours, started by two farm women, Victoria Lipovsky & Kelly Kahman, who create local tour packages in their own Nebraska region, which include local farms. The sidebar, "Regional Flavor," offers a study of the

The Skagitonians to Preserve Farmland work with local farmers to put up crop identification signs, a program varioius other locations are finding effective as well.

Larger and/or long-established entities sometimes have huge numbers of contacts and connections to people who might like to know about your farm and its products and services. When becoming part of their team, the shared promotion and advertising for your farm can reach many more people than you might otherwise if you only promoted independently.

benefits of farms networking with various other businesses to establish a recognizable, yet changing and evolving, regional flavor. An example of regional farming advocacy groups is Oregon Tilth which promotes sustainable and alternative farms in their area by holding eco-farm fairs and publishing literature on Oregon's sustainable farms.

Jan Joannides offers another example of a local or regional entity that helps local agritourism farms. Joan is the executive director and co-founder of Renewing the Countryside, a project to help restore the authenticity and vitality of rural areas. When asked about networking with others as a way to enhance promotion of agritourism farms, she states, "I think some of the most successful examples have been when a cluster of farms co-market." She has helped launch Green Routes in Minnesota, which she describes as "helping travelers find cool, authentic places to visit—including farms."

Local arts festival committees may be another group for you to consider. On Lummi Island, another secluded Pacific Northwest island, numerous artists work and live, enjoying their island hideaway. But twice a year, they organize and participate in an artists' studio open house and self-guided tour, which helps potential customers discover the works of these otherwise little-known artists and craftspeople. Recently, they began allowing the local farms to be part of the tour. Tree Frog Farm, Inc., which produces native nursery plants and sells value-added

CHAPTER ELEVEN

Vendor selling handcrafted goods at the Lummi Island self-guided tour of local farms and artists.

healing products from their farm on the island, participates in this tour and sees numerous visitors come to the farm because of it.

Some festival planners prefer to stay very pure to their theme. The Skagit Valley Farm Tour does not allow craft vendors or musicians, and in the same manner, some art related tours and festivals may not want farm related themes as part of their venue. So, don't take it personally if "farms" are disallowed. They just want to stick to a specific theme. However, the addition of local farms to the art studio tours on Lummi Island made the tour more attractive with more options for those who drove the distance and crossed the ferryboat to get there.

Consider, also, shared interest groups as a place for networking: in other words, groups that have an indirect interest in your farm: A handspinners' group if you own a wool ranch, a sustainable living group if your farm operates with sustainable practices. "People are increasingly interested in authentic— aka real—experiences," says John Ivanko of Green Country, Wisconsin, USA. His Inn Serendipity Farm and Bed & Breakfast provides farmstay experiences of abundant crops and wildlife, flower fields, roaming chickens, solar panels and other renewable energy stations that power the farm, and fresh, seasonal home-cooked meals with organic ingredients harvested from a hundred feet of the back door. "Most of our customers hear about us through the same organizations we support, like the Midwest Renewable Energy Association or at sustainable living fairs or organic farming conferences."

What are the benefits?

Larger and/or long-established entities sometimes have huge numbers of contacts and connections to people who might like to know about your farm and its products and services. When becoming part of their team, the shared promotion and advertising for your farm can reach many more people than you might otherwise if you only promoted independently. For example, the Skagit Valley Farm Tour prints beautiful, full-color literature describing each farm on the tour—paid for by donations offered to the group—with maps leading to each farm, and handles questions and concerns through their website. The promotional materials are distributed far and wide before the tour. Sometimes, larger crowds can be drawn when there are many choices from which to choose, rather than just one (i.e., the "mall effect"). People will bring their families "to where the action is" when they wouldn't otherwise, upon seeing the energy behind an event's promotion. Then, they will see what really goes on at your farm: your products, on-farm activities, and perhaps other events that are well worth additional visits in the future.

What to watch out for

Watch out for larger entities that may want to go too far in defining what and who a "farm" is supposed to be in order to be listed or part of their program, and those that don't seem to be concerned that your farm benefits financially now and down the road from their program. As you will read about in the sidebar, Regional Flavor, the Association for Enterprise Opportunity conducted a lengthy study on what makes economy work, and a rural business' individual uniqueness and ability to change and grow as the owner chooses are major players in the success formula they uncovered. Anything less will eventually backfire. Definitely be alarmed if any entity wants you to promote or be listed exclusively through them, and wants you to offer very stiff discounts to agritourism visitors that come through their entity to you. This cuts you off from all the other methods of attracting customers via other routes for the price only you know needs to be collected to make agritourism in any given year pay.

Also, some larger entities may not attract the right type of folks to your farm. If you have a website for your farm, learn how to check where the sources of those contacting you via your website are coming from. If your site is linked to a website, for instance, listing national festivals, and you find you're getting many quality visitors to your agritourism offering from this listing, you'll know to stick with this listing. If, on the other hand, you're swamped with offers to sell you items for promoting your agritourism, or from more vendors wanting to sell at your farm's event that you'd prefer to have to deal with, you may find that the readers of that website aren't

the quality, potential farm customers seeking agritourism adventures you're looking for, and you may opt to pull your site off.

This isn't to inspire suspicion because an agritourism entity is large. They see the bigger picture, and may have some very good insights. They may insist that your $400 per night for a family to camp in a tent on your farm and enjoy the farm sights and activities is overpriced, or perhaps they'll have good ideas on how to turn $40 per night closer to $400 with some additional farm attractions that won't set you back moneywise or timewise.

For example, agritourism farms can check similar tourist opportunities to help them choose their prices. But this isn't foolproof and can be difficult. A farmer may want to offer a tent camping B&B similar to MaryJane's Farm (see Chapter 3). This farmer's local state parks may charge $12 a night for camping spaces that offer bath houses, a swimming lake, horseshoes, and a private outdoor barbecue, so he may think he needs to charge close to $12 per night. But a larger agritourism group wanting your farm's tourism program to succeed may be aware of Wilderness Outpost, an eco tent-camping destination in Canada that offers local foods and outdoor activities in its temperate rainforest location, and charges customers $4,700 Canadian per person for a three day minimum, and up to $9,450 per person for seven days. A larger entity used to dealing with a diversity of successful tourist programs may help you become aware of different types of potential customers, and also differences and similarities between the low extremes of the state parks and the high-end destinations such as Wilderness Outpost, and help you come up with a price that reflects your unique offering, and helps you profit while attracting the right customers willing to pay for your farmstay. MaryJane's Farm at this writing charges $169 per night for a two-night weekend minimum stay in their wall tent B&B.

Depending on your country, many feel you can trust most non-profit groups with a stated intention to help farmers. And there's certainly nothing automatically unethical about a for-profit business, either, set up to help farms or other earth-friendly enterprises. The Sustainable Energy in Motion Bicycle Tour out of Oregon, USA, has blended business with education and vacations that are good for the vacationer and the health of the planet and local community. They take people on bicycle tours of Oregon's organic farms and intentional communities to learn about sustainable farming, permaculture, green-building, small-scale economies and sustainable energy.

But Nikki Rose, founder and director of Crete's Culinary Sanctuaries, has seen a different story, which I invited her to share. Nikki, who was mentioned in Chapter 1 and further in the epilogue, organizes tours of authentic Cretan culture, including their farms. She has witnessed the problem of large, for-profit entities trying to take advantage of local farmers who are unaware of their farms' importance and value in the world of eco-tourism. Cretan farmers have been described as the originators of

Organized bicycle tour arranged stops along the way at local farms.

true organic farming practices, and many are operating just as they have for centuries, growing and grinding their own wheat, harvesting wild oregano and other indigenous herbs, producing organic olive oil and cheese from their own groves and livestock. She is concerned about big business in destination hot spots such as Greece and Crete. These include the large travel resorts that see eco-tourism as a commodity, the latest hot trend—or as Nikki has described it, "that fuzzy, rural tourism trend they see but have no real compassion for."

Such opportunistic businesses use local Cretan farmers as lures to bring in high-paying tourists, but something also has to be done to help the farmers benefit from this exposure, whether that be support for their own agritourism for which they charge the tourists, or some form of compensation. If not, subsequently, down the road, cost of living in the area grows because of all the high-end travelers and second-home owners the farmers draw in, but the farmers' salary stays the same, eventually closing down the original local farmers. Nikki operates Crete's Culinary Sanctuaries, however, as an internationally acclaimed sustainable travel program that is actively working on projects to help preserve Crete's cultural heritage and environment. She makes sure visitors are directly contributing to those projects and are helping to continue their work.

Here's what Nikki has to say to farmers and others who may be involved in agritourism in similar situations where indigenous, generational local farmers are being approached by out-of-the-area big business, or even their own governments which do not appear to have sustainable foresight.

"Farmers should research and understand how the tourism industry works… and how it might work for them," Nikki says. She is in agreement with many that farms can't thrive if they remain an underclass expecting a larger authority to handle their destiny. Some of the tour operators Nikki has been involved with are in the business of making money at all costs, regardless of the consequences to the environment or local culture. They follow the trends in tourism and a great percentage of them have no compassion whatsoever about the people that provide the content for their customers. "The farmers, chefs, and fishermen make this whole new (eco-tourism) trend happen for tour operators that know nothing about farming," Nikki says. "Farmers should look at examples of how successful agritourism operations work and ask the owners and farmers directly for insight on the pros and cons of hosting visitors… there are plenty of each!"

"Farmers need to know their own financial worth in this scheme," Nikki continues. "Without farmers… there is no agritourism. So if travel agents and local governments are shoving agritourism down local farmers' throats and paying them next-to-nothing to offer these services to visitors, they need to know what the going rate might be for their services and stick to that rate. If there is no going rate, they should look at what the fee tour operators are charging for the whole trip and decide how much they deserve to make by offering their precious time and services. Then they should probably double it because farmers are always too modest about their skills and how much they are worth to others."

Nikki, who has seen numerous problem situations in various countries, feels that farmers should work collectively on agritourism. "They need to know how to put a price on their services offered and stick to it… collectively. All farmers in the region should get together and decide on hourly rates for their services offered… just like lawyers and accountants do. It's a business. If they do not all work together, then the travel agents will find one farmer in the region to exploit until he/she wises up and realizes that they are working for some big, mega-buck generating tour operator for nothing.

There's no balance in that at all. Tour operators always expect everyone that provides their services to work for free… that's their mindset. Farmers need to negotiate fees, provide contracts (simple one-page what we will offer, when, how and for how much). Otherwise, tour operators might take that opportunity to walk all over them and possibly never pay them for their services.

"What farmers need to know," Nikki says, "is that they might become the backbone of cultural tourism… since farming and food IS the root of cultural tourism in many instances. Farmers need to know that they have the power to direct and control how tourism is offered and implemented in their region… and make it work to their advantage. The bottom line is that they not only want tourism in their region, but that they are part of the planning from the onset… and that they do not fall victim to schemes, whether private or governmental. If they wait for the government to plan and implement agritourism (which they know nothing about… let's face it), then they either become a) a hideous Disney-style novelty in their region, or b) victims of a cheap holiday route planned by bureaucrats that will never provide supplemental income that's worth the farmer's own investment and time. Farmers need to be a part of the marketing plan as well and create their own marketing plans (whether that is just collaborating with local hotels and other shops or something grander like creating a website)."

Nikki warns that farmers could end up investing in building renovation, brochures, and so on, but never see a visitor if they leave the promotion only in the hands of authorities that know little about farming. Nikki further warns of bureaucrats and foreign advisors coming onto the property of local farms and making suggestions that might be costly or make no sense whatsoever. I'll quickly inject here my own example also mentioned in another chapter. All farms don't have to provide hay rides. Hay rides were originally the result of haying season. If you don't grow or use hay, but would prefer offering candle-making workshops using your farm's flowers and herbs instead, hay rides may be a worthless investment that only serves to make your farm more generic rather than unique.

"Farmers should seriously consider what investments can be made and why," Nikki says. "If the government is offering funding for such investments, farmers also should ask about long-term advisory assistance in developing and marketing their operation. So many times, I've seen the Greek government and European Union fork out cash to farmers to build apartment buildings on their farms… then leave them to run the operation somehow. The percentage of failures is high. Tourism is a specialized trade… it's not enough for a great farmer to have rooms to let and a wonderful cafe on his or her property… they also need the fundamental resources to run a successful agritourism operation that their family can carry on for generations to come. If they don't have that, then they truly are pawns in a quick-fix government scheme to support rural development, which actually backfires."

The main thing to remember to make for-profits and government assistance work for you is, as Nikki advises, to remain in charge. Find strength by networking with other farmers involved in agritourism, as well as small independent business owners in your area, and see every larger entity as something brand new to be considered, rather than automatically stereotyping them as either on the completely good or the completely bad end of a polarity. For-profit businesses, such as the Sustainable Energy in Motion Bicycle Tour mentioned above, can work well with agritourism farms. And as far as government assistance, Indian Chimney Farm in New York benefited well from a government grant to upgrade their farm for agritourism. With this grant, they widened their laneway, added parking spots, fixed their drainage, extended the lane to their barn, and raised their farm sign so everyone driving down the road can now see it. Numerous agritourism farms in the USA have had very good luck with Chambers of Commerce promotion, Department of Agriculture listings, and for-profit Bed and Breakfast Association directories when approached via the farmer's awareness of the tourist industry.

But even if you choose to promote your farm completely independent of established and larger

Nate O'Neil, owner of Frog Song Farm, sets up a farm stand and tasting booth for visiting members of a Slow Food Convivium.

entities, it's good to at least socially network, or read news on the topic, to know what's going on around you. As you grow as an agritourism farm, your choices need to reflect awareness of humankind's direction, avoiding falling into local or regional traps that other eco-tourism projects may fall victim.

Also, you can create your own "larger entity" in a less formal manner. Collaborate with a few others with unique, locally-flavored products or services to market. If your agritourism project offers an historic barn and carriage house offering quilting and other rural craft workshops, connect it with a gourmet restaurant that serves your local, organic produce, an antique shop, and a quilt-supply shop. All four of your businesses can be easily attracting customers who would love to patronize the others in the group. Share cross-promotion and create promotional packages together. Choose your network wisely—there is strength in numbers, and you don't have to struggle in your business alone.

The Slow Food Movement: A close-up of an ally to the agritourism farm

As my husband and I rumbled towards our 7:00 p.m. appointment in our pickup truck, I sliced cheese in my lap, and told him how silky the cheese felt. What does that have to do with local farm promotion? A lot. This cheese was a one-of-a-kind, locally-produced artisan cheese we chose to purchase instead of stopping off at a fast food restaurant. Thanks to a trend called the Slow Food Movement, many more customers of local sustainable farms may be giving their dollars to these farms, now that they've been inspired to slow down and savor sustainably produced food as nature intended food and flavor to be. The Slow Food Movement is spreading globally, and the U.S. office describes itself as: "an educational organization dedicated to promoting stewardship of the land and ecologically sound food production; reviving the kitchen and the table as the centers of pleasure, culture, and community; invigorating and proliferating regional, seasonal culinary traditions;

creating a collaborative, ecologically-oriented, and virtuous globalization; and living a slower and more harmonious rhythm of life."

Started by a northern Italian food journalist in response to the arrival of McDonald's on the steps of Rome, Slow Food is now an international organization with close to 83,000 members in more than 100 countries. It's grown rapidly in the U.S. in the last four years. The journalist, Carlo Petrini, was apparently disgusted with the loss of food's original ability to keep people close to the earth, in touch with artistic inspiration, and bonded to each other. The movement embraces the restoration of a feeling when the mind was quiet enough for cuisine to be inspired by the local farmers and home gardeners: All—at least for some moments in time—was right with the world. Relax. Savor.

The Slow Food Movement has been a definite ally in encouraging local citizens and tourists alike to visit agritourism farms and rediscover the good life that can be found by supporting local farms. Today's Slow Food revival reaches out to consumers and reminds them of the delights of eating locally, seasonally, and enjoying the process of the food's creation. Consumers are reached via taste-testings, publications, festivals, articles, and awards events. Farmers can receive newsletters, magazines, and get directly involved in a number of ways.

For example, there are local Slow Food gatherings, called conviviums, and agritourism farms can check to see if any are near them. "I had done apple varietal tastings and cider events with Slow Food groups in Virginia and New Hampshire previously," says Michael Phillips, author of *The Apple Grower* and owner of Heartsong Farm in New Hampshire along with his wife, Nancy, who is author of *The Herbalist's Way*—previously entitled *The Village Herbalist.* "I also had given my input into the listing of heritage apple varieties in the Ark." But Michael received a phone call to find out he was even more involved in the movement than he'd previously thought.

Even though a very small ferry boat must taxi people to the remote Lummi Island, this island's farmers attract farm customers by giving agritourism tours to professional chefs (see Chapter 15) and by teaming up with the local artists to promote an open studio/open farm self-guided tour twice a year.

The "Ark" that Michael mentioned above is an international project of the Slow Food Movement to "rediscover, catalog, describe and publicize forgotten flavors. It is a metaphorical recipient of excellent gastronomic products that are threatened by industrial standardization, hygiene laws, the regulations of large-scale distribution and environmental damage. Ark products range from the Italian Valchiavenna goat to the American Navajo-Churro sheep, from the last indigenous Irish cattle breed, the Kerry, to a unique variety of Greek fava beans grown only on the island of Santorini. All are endangered products that have real economic viability and commercial potential."

In the USA, the Slow Food Movement has recently created the Betsy Lydon Slow Food Ark Award, which recognizes one farmer each year who works earnestly to supply wholesome food to his or her local community and inspires others to do the same. Michael was the first recipient of this award. "The Betsy award began with an entirely surprising phone call from the New York City group headed up by Jeffrey Lydon and Hilary Baum," Michael says. "They desired the first award to go to an apple grower because of Betsy Lydon's work with Core Values (a marketing program to give locally-grown

fruit distinction). My name was recommended to the committee primarily because of my book and continual speaking outreach."

With more and more of the mainstream coming to recognize the idea of Slow Food, even restaurants are boasting Slow Food and making an issue of purchases from local sustainable farms, creating a win-win situation for themselves and their farm neighbors. At Mount Bakery, a popular Belgian bakery café in Bellingham, Washington, the owner credits local farms right on his menu, mentioning the local Fairhaven Flour Mill and K&M River Farm. And on the bakery's window, below its name, the words "Slow Food" are painted.

Slow food promotion reminds people of heirloom vegetables like violet-fleshed potatoes. It celebrates farm-processed jam from handpicked fall-bearing raspberries and baked goods filled with wild huckleberries gathered from nearby mountains. It embraces private-label organic cider tours, raw local honey sold at roadside stands, and the silky cheese my husband, Kipp and I, enjoy tonight as we drive. The cheese, named "Mont Blanchard" after a local mountain, was turned daily, aged five months, and then purchased by us directly from Samish Bay

Regional Flavor: How agritourism can network with others to help create solid new economies.

A recent study by the Association for Enterprise Opportunity (AEO) resulted in a publication to help rural and microenterprises succeed and become catalysts for building regional economies. This publication, as well as the association itself, is a treasury of discoveries and insights for a wonderful new synergy among agritourism farms, artisans, and sometimes even urban enterprises. Although its title is Regional Flavor: Marketing Rural America's Unique Assets, much of its material is universal for any country, allowing regions everywhere to support individualism while using local flavor to become the backbone of strong nations and continents. The information was based on a recent project involving "Rural Micro-enterprise Learning Clusters," and aims to restore, enhance and support private innovation and individual entrepreneurialship while crossing past boundaries from rural to urban, and from government to private citizen.

Here are some if the highlights of its regional flavor strategies that any entity working with agri-tourism farms, or any farmers becoming involved with a larger entity, should consider:

"Through a Regional Flavor strategy, rural development work is tied together across sectors, geographic boundaries, and other divisions, accelerating the growth of new economic opportunities. 'Flavor' refers to the variety of home grown ingredients (or attributes), all combining to a single recipe (or experience) where each ingredient still makes a unique contribution. The area becomes known for this exciting, organic identity where a person can have authentic and varied experiences.

"People who experience the one-of-a-kind characteristics of an area are more likely to build an emotional bond to the area, which increases repeat visits and purchases, as well as referrals to friends and neighbors—all resulting in greater economic activity.

"When a regional flavor strategy as described in this study is used, it simultaneously makes the region more enjoyable, accessible, and home-like for its own residents, a top priority.

"All locally owned businesses should be encouraged and sup-ported in creating their own unique character and to innovate. The fact that they will grow and change according to their own inner callings should be a part of the overall equation.

"Any larger entity should help the local business draw on what's unique about the business, rather than pressuring for homogenization.

"Visitors link their target interest in an area, such as bird watching, to other regional attractions once they've come to the region. One business can help bring in business for another without competition."

The publication covers aspects such as the fundamental steps for operating regionally and mobilizing assets, supporting entrepreneurs, removing barriers to increase opportunity, and getting the word out. For more information on the Association for Enterprise Opportunity, or to order this publication, write to AEO, 1601 Kent Street, Suite 1101, Arlington, VA 22209.

Or see their website at:
www.microenterpriseworks.org

In area where Slow Food and quince farm are described. Caption: Group of Slow Food members enjoys tour on an organic quince farm.

Cheese created on Rootabaga Country Farm in Bow, Washington.

The farm is owned by Roger & Suzanne Wechsler. "We've followed the Slow Food Movement for a long time," Suzanne says. More and more eco-farmers I interview are like that: quietly well aware of positive trends before mainstream slickers know what happened. On Rootabaga Country Farm, the cows actually go into the sunshine, consume green grass, and the resulting artisan cheeses are sold directly to eager local customers and lucky tourists. With this book deadline looming, I didn't get a chance to interview the cows. But I had my questions ready: Did you immediately remember what to do with real grass, or were you put through a rehabilitation of sorts? For example, were you exposed little by little to live chlorophyll, fresh air, breeze, nature-given rhythms of lowering your heads to the green earth and then relaxing in the fields to properly digest and create milk as nature intended? How did you first handle the full-spectrum sunrays warming your bodies and the healing, high frequencies of birdsong swirling around your auditory ca-

nals? How long did it take the farm therapist to re-teach you to chew your cuds?

I can only guess at their answers, but if it turns out the cows produce better, have less disease, and remember the good life without therapists, could we? Could the very thing we rush from, the slow life, be something we're innately created for, and something that actually can make us more productive? Do humans remember this, even if the memory is temporarily forgotten amidst a lot of commercialized noise? Because, if all they need to do is be reminded, rather than manipulated and coerced by aggressive guilt-inducing commercials, that could mean that sustainable, local farming is a great business to be in. It's as though Slow Food asks, would anyone like to remember the flavor described as: "All is right with the world. Relax. Savor?" And it's as though, a growing number of people are responding, "Yes!" Because that flavor seems to seep into the taste of foods grown naturally, created slowly, and consumed lovingly. Somehow, regardless of religion, political viewpoint, income, or race, humans remember that flavor when exposed to it. And they want more.

Those involved in the local food revival, including agritourism food and drink-producing farms, may want to contact local Slow Food offices to see if there's anything they can participate in. The Slow Food Movement has international headquarters in Italy. From there it branches out. There are national offices in other countries including the USA, which then are broken down into regional and local subgroups called "convivia" (or "condotte" in Italy), so that the Movement can be interpreted and supported locally. The Movement says that the leaders of these local/regional convivia organize food and on-farm events and initiatives, raise the profile of products and promote local food artisans and wine cellars. They also organize tasting courses and taste workshops and spread the word of new food develop-

Chapter Resources

Sustainable Energy in Motion Bicycle Tour
www.democracybike.com

Crete's Culinary Sanctuaries
www.cookingincrete.com

Renewing the Countryside
www.renewingthecountryside.org

Inn Serendipity
www.innserendipity.com

Online centers where your farm may qualify for a listing (listings are often free).

www.agritourismworld.com

www.biodynamics.com

www.chefscollaborative.org

www.farmerchefconnection.org

www.farmstop.com

www.festivalsandevents.com

www.foodroutes.org

www.landstewardshipproject.org

www.localharvest.com

---

ments, newly rediscovered heirlooms, and knowledge of the products and cuisines of other areas.

Humankind is looking for something that hasn't been around for a long time, and many local farmers are providing it. This "something" is sometimes hard to describe, but it's recognized when people drive by U-pick blueberry patches, seeing customers sitting in the sun, placing each ripened gem gently into their pails or baskets. How can they do that: Work towards pie or jam in such a slow manner instead of doing something more pressing—unless they know a secret… that all is well with the world? They can relax. They can savor. Sure, it's the berries. But it's also that secret customers are after. ❧

Signs & Curb Appeal   12

Whether your agritourism consists of a seasonal roadside stand or a destination farm B&B attracting international guests, curb appeal and well-chosen and placed signs can add personality, and even elegance, while attracting attention and showing the way. Or, they can clutter and distract, and the signs can be nearly impossible to read.

Road signs are often the first impressions customers have of your agritourism farm, so make sure your signs are professional-looking, neat and clean. According to Eric Gibson, author of *Sell What You Sow!* "The road sign identifies your business and directs customers to your farm. A good sign can be relatively inexpensive, yet works for you on a daily basis. A professionally made sign, 4-by-8 feet, costing about $600 and lasting five years, costs only 33 cents a day!"

Size and location—what to consider

Signs have been with us since the dawn of time. Early humans painted and chiseled their territories, hopes and triumphs on stones, creating primitive signs that advertised their cultures. Through such Stone Age signs, fellow humans knew what to expect in a given location. Your agritourism sign will do just that, alert fellow humans as to what to expect on your farm, and should reflect your farm's personality. Its location also should help it be readable. "The first thing I look at when making recommendations about the location of signage," says Kipp Davis, a 30-year professional sign designer, "is: How is the audience, the customer, going to approach the site? They need to have time to read the sign, make a decision to stop, slow down, and make the turn into the farm or roadside stand."

For agritourism enterprises such as on-farm stores, U-picks and roadside stands that need to continually attract spontaneous customers, Eric suggests placing an advance road sign two to ten miles from your farm (one in each direction) to alert motorists when to start looking for your market. Place another sign one-half mile from your farm, with perhaps another sign one-quarter mile away, and your best sign at the farm itself. (Be sure to include the words Next Exit where appropriate!) Too many signs, too closely spaced, only confuse the driver and cheapen the image of your farm's market. Road signs also can be used to show the dates and hours you are open; a customer who has driven several miles only to find your market closed is likely to be perturbed!"

Other agritourism destinations may want less signage attention from passersby, such as the farm B&B which needs its customers who already have been booked online to find it easily, and perhaps could use a little exposure to passing drivers, yet wants to maintain more privacy for itself and guests without inviting unplanned drop-ins.

Either way, Kipp adds that we must think about what it's like to be a new or infrequent driver on our

Signs have been with us since the dawn of time. Early humans painted and chiseled their territories, hopes and triumphs on stones, creating primitive signs that advertised their cultures. Through such Stone Age signs, fellow humans knew what to expect in a given location. Your agritourism sign will do just that, alert fellow humans as to what to expect on your farm, and should reflect your farm's personality.

own road when choosing the sign's location. This can be difficult—perhaps because we know our own routes too well—and we can't feel what it would be like to drive through our own territory for the first time. And, even though a brand new sign might stand out to us because it's something new to our familiar territory, others may be looking at many things anew on our road.

If planning just one main entry sign to your farm, the size of the sign, of course, will also be a factor in choosing its location. "On a roadway," Kipp says, "the speed drivers are going may determine the size of the sign which, in turn, will have a bearing on where you can put the sign. If you want people to see it from two directions, you will need to decide between a one-sided sign that faces parallel to the street, or a two-sided sign on posts that can be seen from either direction. You want to put it in a place that won't block a view or hide a driveway. You may want your main sign to give advance warning as your customers come around a curve or have additional signs at a crossroad, pointing the direction to the place."

Back when we sold lemonade with our nine-year-old friends, our illegible crayon sign that flopped in the wind most likely wasn't what drew in customers. It was probably the larger image of chil-dren sitting behind a card table, obviously wanting to sell something. However, a 14-year-old in Seattle, Washington sold thousands of dollars of lemonade one summer at a busy park with the help of signs that were a little more sophisticated. She put easily read signs up a quarter of a mile before her lemonade stand, alerting joggers, bicyclists and roller skaters.

Regulations

Another consideration, of course, is the regulations in your location for putting up signs. The rules that continually seem to thicken our country's rule books, including those about signage, can be overwhelming. But a closer look shows that many rules are simply common sense, and have to be put in the books because too many people don't follow common sense when erecting structures, and authorities have had to be called in. For example, even though a sign in the middle of the road would certainly attract attention, we can't put one there. Nor can we put a sign on our neighbor's property without their permission, even though more drivers might be alerted sooner as they come around the corner towards our property. Since, again, we're sometimes a little too familiar with our own roadways, driving home by instinct, we may not be as aware that a sign might block a new drivers' view for pulling out of the property or seeing traffic up ahead. Therefore, even signs on our own property may have restrictions. Also, too many signs clutter our roads and distract drivers from directional signs and speed limit signs, so signs put up on county roads or state highways may need permission, may need to follow design and size rules, and may also be taxed. You may be able to strike a bargain with your neighbor by putting up a sign on his or her property for a small fee. To make sure you're complying with local regulations, even with signs on your own property, contact local authorities in your area, which may include your state and county authorities, such as the County Planning Department and Public Works Department.

CHAPTER TWELVE

This sign welcomes visitors to MaryJanesFarm.

Designing, style and practical materials

"Signs must be easy to read, so keep the message short," says Eric. "Six words is about all people can comprehend while whizzing by on the highway." Focus wording and images on what appeals most to your customers. For example, are your customers only pre-registered, overnight guests? In that case, they may like the appeal of one elegant, smaller entrance sign that lets them know they've arrived, but without the fanfare of many signs along the way, giving the feel of having found something somewhat secluded or even secret. On the other hand, when attempting to draw in many ongoing and spontaneous customers such as with a roadside stand, Eric suggests focusing signs on what would appeal to those customers, such as a seasonal product you are featuring. In this case, Eric says, "Use selling words like fresh, homegrown, organic, etc. Periodically change some of the copy or design element on your signs to create continued interest."

A creative sign makes a huge statement. From the first instant people see an entrance, they will begin forming an image envisioning what lies ahead. Some creative and effective signs I've seen include signs mounted on antique farm equipment, temporary signs secured within a wheelbarrow full of flowers, and those that resemble elegant, brass, equestrian farm or country estate signs. I've also seen signs that, while quite pretty, can't be read while driving by. Once, when passing such a sign, out of curiosity I turned around, stopped, squinted, and finally read what the sign says, but most passersby won't take the time. As Eric points out, passersby's safety is another consideration. "Help them drive safely by not having to squint or turn around to read your sign!"

Readability is as important, if not more important, than ornamentation. Make sure the letters are large enough that drivers can read your signs. "Allow a minimum of one-inch of letter size for each ten miles per hour of the speed of traffic passing your site," says Eric. "In other words, if traffic goes 40 m.p.h., use four-inch letters or larger. Test your sign's legibility by driving past your business (perhaps with someone who doesn't already know what your sign says) and see if they can read it from a distance."

For temporary signs, I've seen very effective A-frames set out, similar to the hinged A-frame signs set out by realtors when announcing an open house nearby. Kipp warns, however, that portable signs are prone to being re-arranged by the wind, as well as by uninvited human intervention.

Size

Again, think about your potential customers when deciding on the size, whether you'll have one entrance sign, or several. "If your customers are primarily transient," Eric says, "you might want to make your signs large to attract their attention. But if most of your customers are local and are aware of your market, smaller signs that merely remind them you are open, and give news about new or in-season products, may be sufficient. Also, if a small sign would be overshadowed by others near it, it is probably advisable to use a larger sign if you can afford it."

Materials

Materials have come a long way since the days of crayon and carving on stone, although I have also seen an effective sign painted on a large boulder nestled at the edge of a farm's property! When asked about most professional signs today, however, Kipp says that many are done on plywood. "But not on ordinary plywood," he says. "MDO (medium density overlay) is a marine-grade plywood, which will last longer outside, has a very smooth surface that holds paint well, and makes it easy to apply lettering and graphics, or illustrations. Another type of sign that makes a big impact would be a 3-dimensional sign, sandblasted, carved or with dimensional letters

An old crate is used to make a very readable temporary sign.

added. That may be made of two inch to three inch thick cedar or redwood."

Although the colors chosen for the sign may seem determined only by their beauty, readability, and how well they fit with what you're sign wants to portray (such as painting letters in lavender for a lavender farm), there is even more to consider, including the material used for the lettering. "Certain colors are more prone to fading than others," Kipp says. "Red probably fades the worst. You will likely still use paint for the background of the sign, but nowadays, sign makers use self-adhesive vinyl lettering that is more durable and light-fast, more accurate, and more economical to produce than the old-fashioned, hand-painted sign. Trained professional sign painters can still use the hands-on skills of their craft, but now have the added tool of the computer to produce lettering and graphics from vinyl with as much or as little detail as they used in the past. For many old-fashioned, professional sign painters, though, it's still more fun to hand paint a sign!"

Handmade, quick sign shop, or professionally, custom-made?

"If you are going to do it yourself, I would still use MDO," Kipp says. "You might be able to get a half or quarter-sheet of it at your local hardware store." Use an enamel paint for the background and a high-contrast color for the lettering—darker letters

(Left) Some farm signs can be extremely simple. Farm visitors appreciate crop signs.

(Right) An A-frame can make an effective temporary sign.

Pumpkins line the walkway for this farm's Fall Festival.

Little touches, such as flowers growing in this old stone farmhouse window in France, can make a big difference in the appeal of the farm for visitors. They should reflect your own personality and what you would enjoy creating, anyway. Small touches subliminally suggest that all is well with the farm, and the farmers are content and doing well enough to be creative and beautify their own surroundings.

An antique piece of farm machinery makes an eye-catching sign.

Easy-to-read words and large images make an attractive farm sign.

CHAPTER TWELVE

This scarecrow at Shuh farms makes a fun attraction for their fall festival.

on a light background or light colors on a dark background. Use a single stroke letter (meaning don't go over it) of an even width. If you're inexperienced, don't try any fancy lettering."

There are also options in between custom-made and doing it yourself. Some companies specialize in farm, estate and garden signs designed after country-style themes or elegant, cast-bronze estate signs. With these companies, you have a choice, for example, of a few sign shapes and sizes, a few lettering styles, and whether or not, for example, you want a strawberry, apple or sheep to decorate the sign. Your town may also have a "quick-sign" shop nearby. Rather than doing it yourself, Kipp feels these shops may make a more readable and durable sign. "I think the best advice I could give (for those not needing anything elaborate) would be to have a cheap sign done at a quick-sign shop," he says. "They can make a simple sign on a PVC-like plastic, fairly inexpensively. Your time is worth money, and

you would spend much more time getting materials and trying to make a decent sign, than the relative cost of having it made for you."

More design tips

Backing up to that story of having to stop, turn around, and squint to read a sign, it is possible to have elegant lettering and decoration, but not to the determent of people passing by without being able to decipher your main message. However, you can have some smaller lettering that gives information and is readable only when the drivers slow down after you've caught their attention with something larger. "When designing a sign," Kipp says. "You want the main point to come across quickly and simply. If you are making a sign that says something such as "Island Meadow Farm, Turn Right Here"... "Farm" might be the most important point, the biggest lettering, right in the center of the sign in a simple, easy to read letter style such as Arial or

Signs can be adorned to reflect the current season.

different elements of the lettering, and illustrations or graphics."

But as far as your sign giving out added information, Eric has a warning, "Do not mention prices. Signs at or near the check-in station are better for showing prices. If your prices are higher than the supermarkets,' customers may not stop, but if they see how fresh and tasty and delightful your corn looks, then notice the higher price, they won t be as reluctant to buy from you." Of course, like all rules, there may be exceptions to the avoidance of putting prices on signs, such as a temporary sign that reads, "Free pumpkin for all shoppers today only."

In some cases, "shapes" in addition to the sign itself, may do the initial attracting, while the actual sign simply gives more secondary details. For example, a huge, smiling, bright-orange inflated pumpkin or banner might be effective at calling attention to your weekend pumpkin patch sale, while a simple, more traditional sign next to it gives the rest of the details once people get closer. But as the old song goes, "Sign, sign, everywhere a sign, blockin' out the scenery, breakin' my mind," some attention-getting schemes also can be a turn-off. Signs that carry the subliminal personality of a screaming salesman in a bright plaid suit, or exploit women or movie stars, or bring neon into the countryside, may attract the wrong kind of attention, and turn away those you really want to invite. If possible, regardless of whether you're selling extra heirloom tomatoes or alerting people to your corn maze, think about the overall property's image. Then allow a sign to become an effective part within that whole which enhances the overall image while doing its job of leading or drawing in people at the same time.

Considering the idea of shapes, the sign itself can be an attention-grabbing shape. One man who sells

Times Roman. "Turn Right Here" would be smaller but simple as well, because it is giving a direction. "Island Meadow" could be a fancier script in a less contrasting color, because it is more like an adjective, describing which farm it is. Again, lettering that is bigger and uses a simpler letter style and has more contrast against its background will be easier to read by potential customers. Some other design elements that will attract attention might include things like a border, sign shape, flags, shading and outline on

CHAPTER TWELVE

hydroponically grown tomatoes has mounted a huge sign the shape of a tomato on the edge of his property where a lot of fast traffic passes. Even though the red in his tomato may need to be repainted often, it certainly stands out and tells people whizzing past exactly what the subject of the sign is. Once people have chosen to go off the main road and down the country road closer to the actual farm buildings, another simpler rectangle sign describes the name of the farm and how the tomatoes are grown, letting drivers know they're reaching the correct destination.

"Plywood cut-outs of fruits or vegetables, or wood with scenic carvings and delicately-shaded colors, can help depict a farm-fresh image," adds Eric. "Consider constructing your road signs with individual panels hanging from the bottom of the main sign, listing the types of produce in season (or the events currently in progress). When strawberries are in season, for example, you can suspend a panel saying, 'Strawberries!' Pictorial symbols such as your farm logo or a fruit or vegetable caricature convey ideas better than words. Choose colors which contrast well with each other. Consider also the image you are trying to convey: reds, yellows and oranges proj-

A planting of wildflowers along the farm's road where visitors drive can add casual and natural curb appeal.

ect warmth, action and excitement, while blues and greens are cool (and calming) colors."

Signs provided by larger organizations

The French have a system of signs along their main freeways that lead both foreign and domestic travelers to their incredible historical tourist destinations such as castles and prehistoric cave dwellings. The signs are simple, large, and have a consistent, recognizable artistic flavor that doesn't distract from the lovely French countryside. Their consistency in design helps drivers instantly recognize what to look for when seeking tourist destinations, rather than trying to decipher a jumble of independently designed and posted signs. In Washington State, USA, a local farm advocacy group organizes crop identification signs, and places attractive signs near farming territory to help promote its local farms. And in Canada, the Agriculture and Lands Minister and Transportation Minister announced a new road sign program to help residents and tourists find certified agritourism destinations. They stated that agritourism was growing and had huge potential, and that the signs would not only direct drivers to this new adventure, but raise public awareness of the region's agricultural industry. If you're involved with a regional farm advocacy group, you may want to take note of these and other successful programs for cooperatively attracting farm customers with signage. Make sure, however, that farming in general is being supported by these signs, rather than excluding or making less visible in comparison all farmers who can't pay high fees to the group, and favoring only those who already have high revenue.

Curb appeal and special touches

Don't become inauthentic, but clean up for the safety and comfort of your guests as you would for Uncle Jim and Aunt Clara when they come to visit. For instance, when a group of Slow Food members attended a farm tour of Skagit River Ranch in Washington State, we walked around farm equipment parked in the field, passed sometimes unidentifiable odds and ends stored for future use, and the bathroom available was an unadorned one usually used for those working the farm. This is what made the farm tour genuine and satisfying—it was a real farm. But, on the other hand, there were no overflowing toilets, garbage cans spilling over with odiferous objects, nor the family's laundry heaped in the farm office where we gathered to begin the tour.

Additional curb appeal and "special touches" can come into the agritourism picture if this is something you have a flair for and reveals you or your farm family's personality and uniqueness. We who live on farms can take for granted all the beauty around us, and ignore the empty planter that we had meant to plant with pansies… but just didn't get around to. Little touches can make a difference, depending on your agritourism design. MaryJane's farm in Idaho, USA, allows visitors to stay in wall tents and use outdoor showers and an outhouse. People come for this rustic experience, not for generic resort luxury. However, the owners also keep fresh cut flowers in each tent, and add little touches here and there that give visual appeal to the experience.

Near La Conner, an artists' and tourists' seaside village in the Pacific Northwest, USA, a farm market sits at the edge of the farm's property. Surrounding the market are the farm's cut-flower gardens, making a stunning reason for drivers to take note and pull over.

Jane Hogue, owner with her husband, Jack, of their seven-acre Prairie Pedlar described in the preface, reports in a recent issue of Progressive Farmer that she actually uses her beautiful hobby gardens as lures to draw in customers to purchase from Prairie Pedlar, which sells rare annuals, perennials and gardening related gifts. Their themed mini-gardens attract people from far and wide, up to 5,000 a year, who often spend money while visiting. The themes include a garden based on Biblical plants, and a garden called the Kinder Garden which is for children; where every letter of the alphabet is represented by a plant with a name that starts with the letter. She states in Progressive Farmer that rural farms can't rely on their neighbors alone to support their businesses, and need to draw customers from a wider area.

But Jane's gardens grew from her newly found passion for gardening when she, a town gal, married her farmer husband. Make sure your signs, curb appeal, and special touches are an extension of what makes you and your farm unique. For both visual appeal and safety for your customers, it can be best to have an outsider, perhaps even an ag-extension agent or other authority, drive to the property and walk your farm for sharp objects or first impressions you may have overlooked. ❧

Part IV: Close Ups

Agritourism with an Educational Emphasis 13

Setting up an on-farm educational program can lead to an enjoyable source of regular added revenue. Many people love having children on the farm, and adult students may come from across the state, or even from across the globe, to learn what you're teaching on the farm. A well planned program can result in a much healthier bottom line with the satisfaction of knowing you're contributing to the preservation of otherwise lost traditional skills and/or important state of the art knowledge to the world. "Our society is mainly an urban society now with little understanding of farm life, cut off from an understanding of our food sources," says Joan Schleh, owner along with her husband and three children of GardenHome Farm. "This leaves us vulnerable to manipulation and propaganda by large pharmaceuticals and agribusinesses. I believe that we must do whatever we can to educate the next generation to be protectors of sustainable living." GardenHome Farm is a micro raw goat milk dairy in the USA Pacific Northwest. Joan's agritourism component is very casual and one of her favorite aspects of the farm. Selected customers can drop by to pick up fresh milk and eggs during the day, and if the family is around and not too busy, they may stop to chat, see the animals, and learn more about sustainable dairying.

Lattin Farms offers a very elaborate example of on-farm education, with detailed programs for school children as well as young adult future farm interns from distant countries. It is a fifth-generation farm that grows raspberries, beans, garlic, onions, summer squash, egg plant, tomatoes, peppers, sweet corn, watermelon, and very sweet cantaloupe in the Lahontan Valley of Fallon, Nevada. Their value-added products made on the farm are many, including their own Roasted Corn Garlic Salsa and Garden Herb Vinaigrette, along with breads, cookies, and other baked goods.

Their first agritourism ventures included an old-fashioned roadside stand set up more than 15 years ago, and a U-pick produce area. The roadside stand was erected across the driveway from their front door. Local and out-of-town visitors eventually came in the hundreds as the reputation of the quality of the farm's produce spread, and the Lattin Farm owners added their value-added products to the produce. Now, they report that tens of thousands of customers return every year. The U-pick is promoted to customers as a way for them to leisurely browse the farm's fruits and vegetables, to pick only what they feel is the best and as much as they want.

But along with five generations of farmers, the Lattin family has also produced five generations of teachers. They have learned how to make educational tours of a farm very beneficial to school children. In this chapter, we'll explore methods for making educational agritourism successful for both child and adult farm students, whether you have an affinity for teaching yourself, or even if education only

Many people love having children on the farm, and adult students may come from across the state, or even from across the globe, to learn what you're teaching on the farm. A well planned program can result in a much healthier bottom line with the satisfaction of knowing you're contributing to the preservation of otherwise lost traditional skills and/or important state of the art knowledge to the world.

reminds you of Mrs. Beastly's "F" on your report about George Washington. To emphasize education for children or adults, you can be involved heavily, even creating lesson plans, or be involved in a very minor way, such as telling kids the names of the different flowers you grow while their teacher ties it all into their ongoing education. Or, you can pretty much just smile at students who arrive and point your finger towards the area where someone else who has partnered with you, such as an herbal workshop facilitator using your farm as its location, is doing all the teaching and splitting the per-head fee with you. But there are definite ways to entice students of all ages to your farm, regardless of your style of interaction.

There is a growing trend to plug farms into the formal educational arena of non-farming citizens, and this agritourism niche can bring a steady supply of customers to your farm. While on-farm educational activities can bring added revenue on its own, educational agritourism that includes children also usually generates loyal customers of your farm products when children describe your farm to their parents.

Just about any type of agritourism encompasses education, of course, including harvest celebrations and U-pick strawberry fields. But here we're focus-

ing on education as a main emphasis. Often, educational agritourism falls into one of the following four categories:

- Learning activities for local and regional public school children;
- Programs for interns—youth to adults—wanting to learn to farm;
- Adult farm-related education which may be as simple as the casual "how we farm" tour given to members of a local Slow Food convivium, to ongoing organized on-farm schools or workshops, such as farm-to-plate cooking schools or country living skills workshops;
- Granting Continuing Education Units (CEUs) which helps students and professionals such as chefs and teachers make the cost of the farm classes worth even more to them.

Some professions requiring licensing or certification require a certain amount of approved CEUs each year for the practitioner to remain in practice. Larger regional chef schools and other educational institutions can help promote your workshops when you offer CEUs approved by their schools. See more about CEUs below.

Education for all ages on topics not necessarily directly related to the farm. In this case, the farm owners may teach the classes themselves or collaborate with other workshop teachers, and the farm's atmosphere enhances the topic being taught. Topics may include landscape artist workshops, writers' retreats, outdoor photography "safaris," or even courses that are somewhat more related to the farm's products, such as blending farm-grown herbs with massage oils for bodywork therapists, or making bridal bouquets with your farm's flowers.

On-farm education for school children

The possibilities for on-farm educational tours and workshops for grade-school students are fun and endless! Many ideas and real life examples have been given in Chapter 2. This author's own farm has held classroom apple pressings and historical re-enactments. Our farm is a short distance from an historical one-room schoolhouse. Children dressed in 19th century clothing and were bused from their public school to our farm to see how horses were harnessed to horse-drawn wagons. Then we all walked from our farm to school (walking to school was an ancient way of transportation that once got children's circulation moving and filled them with fresh air before school started), taking turns in the pony wagon and riding the other hand-led pony. At times, kids with their own riding horses brought them along as well.

As another example, on Quiet Creek Farm in Brooksville, Pennsylvania, owners Rusty and Claire Orner turned their love of farming and teaching into a non-profit educational organization, Quiet Creek Herb Farm & School of Country Living, mentioned in Chapter 5. The non-profit status allowed them to obtain large sums of money to offer in-depth workshops and classes to pre-kindergarten through high school students as well as to other adult teachers. They receive grants and tax-deductible contributions given by government agencies, foundations and individuals, which make their student and teacher workshops possible. Further, they recently were awarded the title of an Educational Improvement Tax Credit Organization. This status allows local corporations to make tax credit contributions towards enhancing private and public education.

At Lattin Farms, the owners realized how the world has changed for children and how disconnected many of them are from the natural world and from agriculture. Their educational program includes the following:

* Lessons on pollination;
* Knowledge about characteristics of animals and plants, including a visit to the farm's special animal viewing area, "Critterville;"

* Measuring methods, harvesting and uses of plants;
* An interactive alfalfa unit to learn the importance of alfalfa in our lives;
* Real farm experience;
* Uses of produce including picking their own produce from the farm's U-pick garden and taste samples of produce prepared in the farm's kitchen;
* A bag of produce or a pumpkin;
* And, of course, the fun memories of experiential learning on a farm.

This bulletin board at a destination lavender farm leads visitors to various educational activities.

Some of the activities included are similar to Lattin Farms' general public farm events, such as entrance to their corn maze and hay rides to the pumpkin patch during pumpkin season. But their classroom educational packages offer more serious lessons along with them.

The tours and workshops include homemade lunch packs along with trips to their country kitchen which offers baked goods and preserves. Students are allowed to bring their own lunch, of course, if they wish.

Reservations are made by phone or faxing an online registration form. Prices per-head depend on which package teachers choose for their classrooms, and self-guided tours are also an option.

This author has worked with the public schools for two decades, and I can share with you that, like everything, the educational world fluctuates between periods of more openness for individual teachers to find creative ways for kids to learn, and periods of stricter control over teachers with emphasis on passing specific national tests. During stricter periods (of which we are in at this writing), you may want to consider one of three options:

1. *Offer a variety of shorter learning field trips for public school teachers to choose from.* These shorter trips don't take as much time away from the children, so shorter programs may be more usable. Teachers today can have overwhelming pressure to make sure students score well on nationally-mandated tests. Children may not have the time, for example, to arrange and participate in a longer on-farm program such as spending a day helping the farmer plant, and returning to the farm several times during the year to participate in the procession of the seasons. They may not even be able to come out for a full day's worth of natural science. Ask teachers if shorter choices would be more valuable. Could they visit the farm for a two-hour bird watching or bug-hunting safari in which students make a list of all the birds or insects seen on the farm to supplement a science unit on these creatures? Or, tour the an-tique-apple orchard for one hour to supplement lessons learned about the cycles of plants, each picking an apple to take with them back to class?

2. *Work directly with teachers to create on-farm activities* that cater to specific tests, perhaps in natural science, history or mathematics. This can lead to longer term and guaranteed programs. For example, two grade levels may partner to have the younger grade plant seeds in the spring and then return as older students the next school year to harvest in fall. The teachers involved, knowing this plan ahead of time, can arrange their own lesson plans on natural science and math, which may include estimating the amount of seeds as compared to projected amount of harvest, weighing final produce, keeping a weather station on the farm, sprouting some of the produce in their own classroom before transplanting on the farm, etc.

3. *Consider also home schooling groups and alternative private schools.* These groups are often more open to teaching children to know what their culture is already like (rather than molding the local culture—or farm—to meet their educational needs). They are often very open to field trips and workshops created by farmers and know how to be flexible to adapt their teaching and lessons around what their local community already has to offer.

During periods of more openness in the public schools, farms can sometimes offer more farmer-teacher created longer-term workshops such as historical re-enactments where students build up to the historical day on the farm in the classroom for sometimes months beforehand, milking to cheese-making, longer term animal care, shearing-to-shawl, planting-to-harvest, and so forth. Here are a few more details to consider:

Begin early when working with specific schools or teachers. Teachers often need plenty of time before their field trip to prepare, and may even need a pre-visit. They may want to plan pre-activities and follow-up activities that merge with their requirements for teaching math and other specific subjects.

Child immersed in the blooming crop of a lavender farm during its annual festival.

They may want to have their students practice new words they'll experience on the farm, or be prepared to count and measure items on the farm, or note ahead of time where the best spot will be to have them all become silent and listen to sounds around them. They need to know what to tell the parents as far as whether rubber boots or long sleeves should be worn, and if there's a covered area to write in their journals.

Take into account the school year schedule. The first week and last week could be too hectic. Kids may all be out on spring break when you'd planned a possible children's spring festival. Whether you plan on publicizing through the school system or having an even closer partnership with specific teachers, they need plenty of lead-time to plan participation. If your program is new, try to allow the teacher(s) to come to the farm before they bring the children. If your time is tight, offer just one weekday teacher's quick open house on a specific day after school hours but before their own family time, such as from 3:30 p.m. to 5 p.m., rather than trying to accommodate every individual teacher's request to come visit at different times.

If you can, involve a teacher who can work from the inside of the local school district and who understands how other teachers will react to your farm's field trip offering, including what information they may need from you first in order to get permission and prepare for the field trip. For example, they may ask advice about what their first aid kits should contain if it's their policy to bring them along, and they may need information from you before checking with their superintendent to see if there is insurance for such field trips, what it will cover, and if not, are special waivers needed for the parents to sign, etc. This is all work the teacher must do, but your answers, "No, there are no horses on the farm; yes, there is an open body of water," etc. will help the teachers prepare. The teacher also can offer valuable insight into effective ways to motivate other teachers to spread the word about your farm's educational opportunuties. Brainstorm with your teacher(s) about participation ideas. As time goes on, keep a collection of these ideas for your own plans or to hand out to new teachers you're enticing to join the farm program, such as:

Assemble a collection of resources for stories to tell about farm life similar to yours, for both before and after the field trip. This would include short story collections, such as those found in *Earth Child* by Kathryn Sheehan and Mary Waidner, Ph.D, which offers "Dream Starter" stories that allow students to imagine being specific animals, which is said to help children understand themselves and their world better by comparing their lives to that of other species on their planet. Story resources also include illustrated picture books, such as *A New Coat for Anna* by Harriet Ziefert, which teaches a good history lesson as a child is taken from raw wool to natural dye to spinning, weaving, a tailor and at last, a bright red new coat.

Pre-visit and post-visit activities that other teachers have done to reinforce the field trip, such as keeping a naturalist's notebook, experiments with plants and weather, putting on plays about farm life, and cooking or making gift crafts from farmed ingredients, sprouting seeds and planting them in a spot on the farm.

For younger children, material suggestions for free-play to imitate farm life after the field trip. James Baldwin is quoted as saying, "Children have never been good at listening to their elders, but they have never failed to imitate them." Other than a very good story that most children love to listen to, young children naturally observe and love to imitate as a form of learning. A collection of items pertaining specifically to your farm's processes, such as pails, harvest baskets, kid-sized gardening tools, seeds, etc., could be something you purchase wholesale and offer as a resale package.

Education for adults

HeartSong Farm in New Hampshire owned by Michael and Nancy Phillips offers classes that are an extension of their own passions and enjoyment of teaching adults who are eager to learn what the Phillips have to offer. They grow antique organic apples and numerous herbs which they harvest for bulk sales and use in value-added products. Their classes include one called "Foundations of Herbal Healing," which they describe in their literature like this:

"We welcome you to join us for a Green adventure that will immerse your heart, mind, and soul in the world of healing herbs. Our program will be an in-depth experience for anyone wanting to use herbs for health and well being. We will identify, grow, harvest, prepare and use herbs together. This course will provide you with the practical skills necessary to become a home herbalist. People well versed in other healing arts find our Foundations program a holistic complement to their own work. The Science and Art of Herbology (a home study course written by Rosemary Gladstar) will be used to augment the hands-on learning here at our farm. You'll be awarded a beautiful certificate at the end of the course to acknowledge your dedication to learning about and honoring the healing plants."

For fun, on-farm adult workshops such as these which may attract out of town students, include, as the Phillips do, a list of links to information on nearby overnight accommodations if the workshops continue for more than a day, and you don't have overnight accommodations for all possible students. Even if you do offer rustic camping, for example, or group sleeping rooms, decide if you're interested in attracting more students, or if the classes are more of an enticement to get more overnight guests. If the former, help those who prefer more private or conventional hotels or other types of overnight stays, which could make a difference in whether or not they'll take your classes. If there will be time for out of town guests to roam your farm, region or nearest town during free time, list the awaiting adventures for them, to help them make the decision that their trip to your workshop will be an enriching one even beyond the workshop itself.

Continuing education units

Continuing Education Units, or CEUs, are a status given to qualified classes and workshops, and usually are presented to the students via a certificate if they complete the course satisfactorily. They are far less formal than college or university credits, but more formal than casually taking a class or being awarded a certificate of completion by the private workshop facilitator. Like college credits, the amount of CEUs one obtains from a class depends on the amount of contact hours the student is involved with, a formula which is explained by the International Association for Continuing Education and Training. Sometimes, students enjoy obtaining CEUs simply for their own satisfaction. CEUs don't replace an actual, accredited, college credit and can never add up to a degree, but are more often thought of as education that continues after the degree or main course of study is complete.

However, certain professions call for those already involved in the profession to get a certain amount of approved CEUs in their field of work (after any degree called for) every year to maintain their license, certification, or otherwise ability to continue practicing. In come cases, adults who are working towards a degree that allows life experience may be able to use CEUs to some extent to reach their goals. Also, CEUs may appear impressive on an application for employment. Plus, once employed, some employers may be willing to pay part or all of the class for his or her employees if CEUs are granted. This has been the case when an organic restaurant pays for their chefs to attend an on-farm culinary school like that of Quillisascut Farm described in Chapter 15. They report that one of their on-farm, four-day courses, "Cultivating Success:

Intro to Farming," is connected to their state's agriculture university and can offer CEUs. For their culinary classes for professionals, they can give certificates with clock hours certified by their own Quillisascut Team, and for students who are instructors, the certificates are honored at most higher education culinary programs. It's good to make sure that any of your classes that offer CEUs or formal, clock-hour certification, also offer an option for students who want to take that same class, but aren't interested in CEU certification. Those who do want CEUs need to fill out more paperwork, let the instructor know ahead of time, and sometimes pay a small additional fee. The others don't have to bother.

Often, CEU classes are offered and put on by organizations such as agricultural extensions, professional associations, and even colleges. Any of these can be contacted if you have an idea for an on-farm class that may be appropriate for CEU-approval by that organization, such as blending herbs for healing massage oils for massage therapists through a nearby

A nice break from electronics and commercialism: two boys enjoy observing the skills of a professional artist invited to Pelindaba Lavender Farm to paint the crop during a farm festival.

massage school, cooking with heirloom vegetables through an established culinary school, or the life cycle of plants to elementary teachers for creating a related lesson plan for their students through their school district. It is suggested you offer the class on a more casual basis first, to get experience and organization skills, and determine the exact amount and hours involved. Then present a well-organized plan for CEUs.

At this writing, the term CEU is reportedly public domain. Anyone can offer them independently. But this freedom has been abused and the prestige of CEUs weakened because of it, so having them sanctioned by a professional association or educational institute where you may gain students is wise. Also, if you want your CEUs sanctioned, but you can't find—or don't want to limit your CEUs to—a specific organization such as a chef's school, there is also the option of having your workshops approved by the International Association for Continuing Education and Training (IACET). This takes longer and involves fees, but carries prestige that may make your workshops better known and more desirable, depending on the amount and type of students you're seeking.

Internships

Academic students and curious travelers alike are seeking farm internships in growing numbers. For sustainable agriculture students, experience in the whole system of the farm is often deemed even more necessary than in conventional agriculture. Unlike the day tour or farm workshop that lasts several hours or is taught on set days of the week, the intern becomes part of the farming team for a committed time frame. In general, intern help is traded for the experience they get from you. Often, basic room and board is provided by the farm. Sometimes, the farmer is paid for their service of teaching; other times, the farmer pays for the high quality of help received from the interns, which is often a lower cost than hiring a professional. Interns may contact you direct-

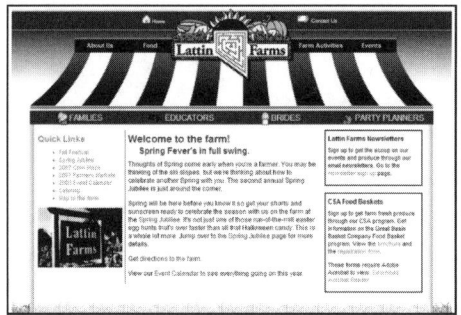

Lattin Farms, www.lattinfarms.com

ly after learning about your farm, or you may find interns through an organization that matches interns with farms. Some organizations simply help set up matches without checking out either the farms or the interns. Others do more "filtering" of the possible interns, and of course, interns that come to your farm through universities might be the most highly filtered of all.

Farmers' experiences with interns vary. Many farmers report that the often youthful and vigorous labor they receive is well worth the time and cost of hosting interns. One USA farmer in who grew hay using only mules and draft horses says he advertised in a regional newspaper each year for interns to stay on the farm and help during haying season. He offered no pay and primitive living conditions, and he attracted good help each year. Another CSA farm found that they occasionally ended up more with the merely curious rather than serious interns, who mysteriously disappeared from their free room and board as soon as they realized actual, often tedious, work was required.

But on Lattin Farms, along with the many school children they serve, hard working interns have stayed on and had life-changing experiences. Here's a quote from one of their interns from Thailand named On-anong Srisuwittanson (On for short). Lauren Augusta provided this to me. Lauren is the Executive Director of MESA, the Multinational Exchange for Agriculture, a 501c3 non-profit and U.S. Department of State-Designated Exchange Vis-

CHAPTER THIRTEEN

itor Training Program. Their slogan is, "Cultivating sustainable farming communities around the world through farmer-to-farmer exchange." Below, I've just slightly edited On's impressive English:

"Here, I am learning more than agriculture. I also learn how all the different pieces of the farm and ag-tourism activities work together to support each other and make something larger. Just like a puzzle. That is how small business keeps running. Doing many things to attract customers and making your own brand name. Let people keep your farm in their mind."

Each fall, Lattin Farms opens their corn maze custom-created by an international designer, and among other activities, allow tours of farm animals and antiques.

"October is the busiest month," On's journal continues. "Lattin Farm got a call from schools every single day to reserve an education tour, maze and pumpkin patch experience. The kids love to come here. They enjoy themselves with all the activities we serve. This month is my favorite month. We were keeping busy all the time. Everything opens full time. Rick told me that this is the month they can make more money, especially from ed-tours and the maze. In my opinion, a corn maze is one way to add value to the corn field. Because you can make more than $1,000 a night from a 3-acre corn field."

As we've seen, educational agritourism ranges from generating revenue directly from the educational activities, to gaining free or lower-cost help from interns, to a huge boost in world of mouth promo-

Chapter Resources

Intern Resources

North America

ATTRA: National Sustainable Agriculture Service. A regularly updated directory of North American farms that take interns also mostly from North America

www.attrainternships.ncat.org

International Association for Continued Education and Training

www.iacet.org

Farm to School

www.farmtoschool.org

USA plus international

WWOOF: World Wide Opportunities on Organic Farms is a cultural exchange.

www.wwoof.org

OrganicVolunteers.com. This site lists your farm by USA state, and has a page linking to WWOOF programs worldwide

www.Organicvolunteers.com

MESA: Multinational Exchange for Sustainable Agriculture

Sponsors farmer-to-farmer exchanges between different countries, matching USA interns to farms in other countries, and matching countries other than USA with USA sustainable farms.

www.mesaprogram.org

You can also list your own farm's internship program independently on your farm's website.

Universities with agricultural departments may also work with farms to set up internships for their students. In some cases, they need their students to observe and help out on one, or a variety, of authentic farms. In other cases, the school may work out a system with the farmer where their students complete an actual beneficial project on the farm previously wasn't already being done by the farmer. Contact your region's university for more information.

Other Resources

GardenHome Farm

www.gardenhomefarm.com

Lattin Farms

www.lattinfarms.com

Quiet Creek Herb Farm

www.quietcreekherbfarm.com

Quillisascut Farm

www.quillisascutcheese.com

tion by those who participate in your education. But farmer-teachers need to be paid for their important work in one way or the other. If you are more formally teaching school children or adults without a fair trade in labor, make sure you charge for the teaching of the wisdom you've gained over the years, and the important contribution you're making to future generations. As mentioned in other chapters, you can always tithe if you feel a need give something away free to the world of education. For every classroom that pays $5 per head (for which the teachers often tell the students' parents to send in before the field trip) offer to pay the cover price for one underprivileged child. If there are more underprivileged children than that, the teacher and his or her school district can better handle that obstacle. Their solution may involve a bake sale where their students earn the money themselves for every student, which can be a very important lesson and self-esteem builder for all students, especially the poor. ⮞

Chapter Resource

Farm-Based Education Association
  www.farmbasededucation.org

Agritourism and the B&B: *You don't have to own a renovated Victorian... Can you pitch a tent?*

14

MaryJane Butters has been there, to that nightmarish brink of possibly losing her farm. But she's a survivor with an entrepreneur's spirit. Ideas sprouted from the ashes and eventually ripened to make her sustainable farm and chosen lifestyle gain ground. One innovation of several is her family's open-air, wall-tent Bed and Breakfast.

MaryJane, an organic farmer in the rolling hills near Moscow, Idaho, says that to help restore farming, we must tell our farm's story, and make our farm a destination. MaryJane was a single mother more than two decades ago, when she bought a five-acre, rundown farm to raise her children. Today, the successful farm's offerings are many, including agritourism. They grow organic fruits and vegetables, produce prepared organic foods and operate an on-farm school. They have also integrated the very unique on-farm bed and breakfast concept which we'll focus on here. Its concepts can be adapted even if you do own a renovated historic farmhouse and are looking for ways to expand beyond its walls, or if you're looking for ways to begin a B&B before actual renovation is complete. Also, their marketing and "special touch" approaches can be utilized by any type of B&B.

The entire farm, operated by MaryJane and her family, now encompasses her initial five acres, along with 50 acres owned by her husband, Nick Ogle. MaryJane was his neighbor before they married in 1993. He left flowers on her doorstep after meeting her for the first time, and the relationship blossomed as they discovered their shared interest in organic farming.

B&Bs today are being celebrated again in new ways, but they are part of the human race's long history. Almost since humankind's beginnings, the B&B concept existed in one form or another. Nomadic tents and monasteries served as B&Bs for travelers, and still do. Although called by many other names, such as minskukus and pousados, the English-speaking term "bed and breakfast" originated in England, Scotland and Ireland. Europe is now covered with an array of farmer-owned B&Bs, showing off everything from the romantic French countryside to British high tea. In the USA, B&Bs date back to the early settlers when pioneers traveled across the country seeking safe refuge in others' homes. During the Great Depression, "boarding house" was the term more often used when homeowners tried to bring in supplemental income by letting rooms and cooking meals. Boarding houses went out of favor after the Depression when large, posh new hotel chains symbolized a new era of prosperity. More often called "tourist homes" for the boarding houses still remaining in the 50s, eventually the charm and advantages of B&Bs returned to American favor. Then eco-, cultural-, historical- and agri-tourism emerged stronger worldwide, and more and more farms once again began to benefit from the B&B concept.

> *Part of the appeal of farm B&Bs worldwide is that no two are alike; they can't be homogenized, and guests enter a world, such as your particular farm, they won't find anywhere else.*

Part of the appeal of farm B&Bs worldwide is that no two are alike; they can't be homogenized, and guests enter a world, such as your particular farm, they won't find anywhere else. Today, historical thatched-roof B&Bs attract visitors in the South African rural garden route. In Wales, Tudors and stone farmhouses welcome guests, and in America, renovated early American Farmhouses and refurbished log cabins attract visitors year-round.

But MaryJane didn't have the income at the time to restore anything. What she had was a farmhouse built by two single brothers in 1905, where she'd raised her kids with no plumbing and an outhouse. But her daughter, Megan, now grown, was about to get married to a young man named Lucas Rae. And from that, the B&B idea was conceived.

Megan tells how it all began: "The B&B sort of evolved on its own. As Lucas and I were planning our wedding, we purchased a couple of wall tents in hopes that our friends and family would enjoy staying at the farm. We got a little carried away with decorating the tents and kept inviting others to stay at the farm. We then realized that we might try a B&B to cover those costs. Slowly," she says, "the B&B idea was formed in hopes of sharing our farm with others and keeping it alive." As time went on, they added a three million dollar liability umbrella to their policy, and the county health department approved their drain fields and outhouses.

Today, it is in full, ever-evolving operation. "We have five secluded wall tents nestled throughout our orchards," says Megan, "each with a salvaged barnwood floor and a full, antique, iron bed blanketed in organic sheets and piled high with goose-down pillows and comforters. The tents are lit with oil lanterns and warmed by wood-burning stoves that also can be used for cooking. Each tent also has front and rear decks for enjoying a good book to the chorus of crickets, and a fully functional outdoor kitchen with propane stove, cold-water sink and campfire. Our guests share our environmentally friendly outhouses and shower houses." Overnight stays include an organic breakfast and an optional mini-harvest from their U-pick garden.

"The B&B has become a fabulous new 'crop' for small farms," Megan says. "We are still working towards our ideals with the B&B, because we are still in the beginning stages. Ideally, the B&B will sustain the farm so we are able to work together as a family and continue to grow and sell an array of organic vegetables. Being able to share those vegetables, foods, and the farm family experience with our B&B guests has been a rather amazing experience. Each family, couple, group of friends, or individual has become a great addition to our farm, whether they stay for one night or the entire week. This week I booked the entire farm for a weekend in June for a family reunion. They are traveling from all over the country to experience the daily routine of the farm."

Could the wall tent idea work for you?

Wall tents are just that—tents with four vertical walls. They can usually be found sold as kits in a variety of sizes, or as canvas and rope only to add to your own frame, or as just the material for the frame for you to add your own cover. Sturdiness varies, of course. Some tents are made for very rugged conditions and are sometimes used by hunters.

MaryJane's farm is situated in Idaho's temperate North American climate. "We have four solid seasons," Megan says after I asked about the climate in which they operate their wall-tent B&B. "During the winter months, we do close the B&B; however the tents stay up year-round. Mostly, (we close down in winter) so that we might enjoy a few months to catch up, and the upkeep of the tents is more work (at that time)—upkeep, meaning tromping through the snow to put clean sheets on the beds and mopping farm mud from the hardwoods. All doable, but we like the weather excuse to catch up on all our other projects. The tents do very well in rainfall and snow. The snow actually acts as an insulator. High winds can cause a restless night because the tent flaps do just that, and flap away through the night. But," she continues, "we remain very dry and warm through each season." Farmers who experience high-wind areas might want to consider locations on their farm within more wind-sheltered microclimates. And the tents on the farm now have wooden shelters over the roofs.

Finding customers for the farm B&B

As far as reaching customers, Megan says that word-of-mouth has been the best method for them. "A simple but well done website with contact information is also a great way to reach a wide array of customers," she says.

Jane Eckert of Eckert Agrimarketing in St. Louis, Missouri, also states that both of these methods work well. "Based on my conversations with B&Bs," she says, "I understand that the Internet is by far their primary source for new business and bookings. I have heard anywhere from 80 – 90 percent of new bookings come online. Once someone has stayed at a property, then the word-of-mouth marketing also becomes very critical. Consequently, a website is critical to the success of building a B&B business."

MaryJane's website also offers an e-newsletter, something Jane Eckert also highly recommends for farm B&B owners. "On-going communications to their best customers via e-newsletters allow them to not only stay in touch but to share changes, new area attractions, etc., thus giving their best customers a reason to return again and again," Jane says.

Authentic farm B&Bs may want to look closely at large tourist entities before affiliating with them as a marketing tool. Ask: do they want to preserve the environment, the

Megan and Lucas test the outdoor camping area for farm guests a week before the season's first visitors will arrive.

diversity and the authenticity of farms, or offer cheap homogenized or fabricated entertainment while they gain profit? The UK, France, and other countries have ethical programs that help promote farm B&Bs, some of which were created by the farmers themselves. In the USA, local Chambers of Commerce may be helpful. "Farm B&Bs should join their state B&B associations," Jane also suggests, "so they can be linked from that state website (which is a primary Google search term—"Kansas bed and breakfast," for example). These state associations often have annual conferences and have qualifications to become a certified establishment that will aid their credibility as a hospitality provider for a prospective guest." If your farm B&B fits the criteria, a state association may be a very useful resource for generating customers. If not (or not yet), numerous farms do succeed in various agritourism operations, including B&Bs, independent of other larger organizations. But networking of some kind to generate word-of-mouth promotion, and website presence, seem to be top priorities.

Farm B&B owners would also benefit from the tips in Chapter 11 on networking with others to create regional flavor and cross promote each other. Besides the farm, the B&B will draw customers

Themes and variations for the new B&B to generate word-of-mouth publicity

Even if you don't plan to hold regular farm events and festivals along with your B&B, a once-in-a-lifetime open house might be a good kick-off for a new B&B to begin generating word-of-mouth. You could simply hold a "new B&B open house" and hope interested visitors arrive, perhaps even participating with your local Home and Garden Tour organization. But adding a theme to the open house may draw more visitors than an open house alone. The possible themes and variations to the traditional B&B described below are well-suited as ongoing themes to many B&Bs, and may also serve as themes for this one-time open house. Chapter 2 gives many other activities you may want to try just once to entice others to see your accommodations. Once you've committed, send a news release to your local media in plenty of time for them to announce it ahead of time to their audience. Be sure to include enticing details that would spark potential visitors' interest, such as any antiques with intriguing histories you use to decorate your B&B. Read the chapter on news releases to make sure yours sounds newsworthy rather than simply an advertisement to drum up business.

♦ Cheese tastings: Doesn't matter if you don't own a dairy, just host an event related to tasting artisan cheeses, ciders, and various pleasures produced by other locals.

♦ Children's birthday or tea parties;

♦ Mystery weekends;

♦ Local author events. Allow a local writers' group to host a variety of local authors and invite the public to come meet them at the farm.

♦ Writers' or artists' retreats or workshops. Open the B&B to facilitators of these workshops.

Your one-time event can lead guests to tell their visiting relatives and out of town friends of a fun, new place they can stay next time they visit. Shared interest artisan groups, such as quilters' groups, often love to go on weekend retreats to rural settings, and your one-time event can begin the rippling effect of these people hearing of you through their social network.

because of their local area's other unique attractions. What we once thought of as just our own boring small town or isolated rural home may fascinate outsiders. Dig into its history, its geology, and find out what unique native plant species and birds live there with you.

The business of B&Bs

There are numerous resources already available for further information on accounting, specific insurance, and other details for those considering a B&B. The Better Business Bureau in the USA, USA extension business start-up plans, and Internet resources for starting a B&B can offer, often free of charge, sample business plans and business checklists online (see contacts below). Suggestions, rules and laws change or become outdated continually, so if concerned, be sure to contact local authorities involved directly about any particular publication's recommendations. Some extension bulletins become unavailable quite quickly as precise legal information becomes out of date. Many books on creating a B&B are general, giving ideas on purchasing an existing B&B or shopping for an urban home to renovate, much of which is of little use to the existing farm wanting to add a B&B within the current farm operation. But they may give ideas on renovation or accounting, even if not necessarily catered to the farm B&B nor the current laws in your area.

As in farming, often the best information for your particular situation comes locally. Find another non-competing bed and breakfast, and ask the right questions on how their particular business works for them. When reading books on B&Bs, it's good to learn what can appeal to customers, and what can bring them back. But don't be intimidated by the rules set out for luxury country inns. These inns are very enjoyable and in some cases may interact with working farms. However, you can choose the type of customer you want to attract, and for many farms, it may not be pampered luxury, but rather nature, a rustic setting, and authentic farm life. In South Africa, my sister non-profit's first food garden planted smack in the middle of a shack town built on an old garbage dump (the sides of the raised garden beds were literally made of old rubbish) are attracting cultural visitors from numerous countries. Start

Wall tent awaiting the arrival of farm guests.

Other alternative forms of guest accommodations

Other alternative forms of guest accommodations beyond the walls of your own home include the list below. Your area's building permit laws and the zoning of your property will determine which are allowable for you. In some locations, rustic camping is allowable as a rural business, and some states have far less restrictions on alternative buildings than others.

Renovated barns or new barns with sleeping lofts.

Adirondack cabins. Boy Scouts may know what these are. Adirondack cabins are very rustic, open-air roofed cabins that allow safe, but full, enjoyment of the great outdoors. Three sides are wood, and the fourth wall is either open, partially screened in, or sometimes open with a primitive fireplace down its center. Inside, there's sometimes a table in the middle and always bunk beds around the edges. Outdoor showers and bathrooms are set up elsewhere. See if your local Boy Scout troupe has information on building Adirondack cabins.

Tiny homes. Tiny homes are a trend for those wanting a separate in-law home or work office, and also can work as farmstay cottages. The books, *A Tiny Home to Call Your Own: Living Well in Just Right Houses* by Patricia Foreman and Andy Lee, and *Tiny Book of Tiny Houses* by Lester R. Walker offer good overviews with many real life ideas. The former includes possibilities for making these tiny homes technically "portable homes" which may have more lenient building code restrictions, and the latter even presenting the inspiration to replicate Henry Thoreau's famous cabin!

Hand-sculpted, earthen cottages. The science of building homes with clay, sand and straw (or cob) has grown immensely. When building this way, one can literally hand-sculpt a home, with storybook, curving walls and gnome-like nooks. The advantages to the farmer may be that much of the materials are already on the farmland, and the building can be worked on slowly over time if necessary. The result will certainly be one-of-a-kind for your guests! *The Cob Builder's Handbook: You Can Hand-Sculpt Your Own Home,* by Becky Lee; and *The Hand-Sculpted House* by Ianto Evans, Michael G. Smith, and Linda Smiley are two excellent books on the topic.

Camping with a twist. Many guests love the pioneer spirit of roughing it in a rustic or historical setting. In England, a woman has set up teepees on her farm for guests to experience the historical lifestyle of native North Americans. Other eco-destinations have set up Old West covered wagons for overnight accomodations, yurts, treehouses, and even a refurbished, rescued boxcar. Some guests like a combination of the wilds along with a little gourmet. MaryJane's Farm has blended the idea of camping in tents with the slightly more adorned B&B. Here's a similar idea from Wilderness Outpost, an eco-destination in the Canadian rainforest. Under a canopy of wild temperate rainforest, guests sleep in tents on wooden foundations with elegant beds and cozy, fluffy bedding, similar to MaryJane's. There are boardwalks from the tents leading to the shower house, an outdoor stone fireplace, a library tent with Internet access, and a spa tent. A bistro tent serves gourmet, regionally grown and harvested meals on elegant china.

with who and what you already are as long as you stay within laws of your area.

You may do well by following these steps:

1. Check local laws about operating a B&B.

2. Add the necessary extra liability coverage and/or change your business structure (for example, from a sole proprietorship to an LLC as explained in Chapter 5.

3. Practice with friends and family.

4. Upgrade current financial farm bookkeeping with the B&B addition; then begin the B&B business in full swing.

5. Use the same good sense in B&B accounting and financing as you do in successful farming. Consider all costs to your B&B, buy wholesale (or second-hand when legally allowed, such as with antiques for decorating) and sell retail. Calculate an overnight price that will bring profit. The exception to buying wholesale and charging retail of course is when offering your own products retail, or offering something one-of-a-kind that repays itself by making your B&B unique. An example of this would be purchasing one-time at full price from regional artisans. If you purchase thick, locally handmade washable quilts for your B&B sleeping quarters from a local quilter (a wonderful deal for such an artisan!), this will make your B&B stand out from the cheap wholesale bedspreads used in massive numbers of rooms at large hotels, and can be financially viable on a B&B farm with just a few rooms.

6. Consider trades; also, asking the quilter for a discount if you help lead your customers to her quilt shop. When consulting other B&B owners about specific how-tos, remember that unless they are in some way legal consultants, they are only telling you their story, not giving you specific legal or financial advice. The final choice and accountability will

Megan Rae prepares flowers for guests in the wall tent at MaryJanesFarm B&B.

Inside a wall tent that serves as a farm B&B at MaryJane's Farm.

There is no cookie-cutter plan for all farm B&Bs. They are all so unique. Mary-JanesFarm B&B evolved on its own without any formal business plan at all. But it did spring from a firm foundation: the family of MaryJanesFarm very much wants to preserve authentic farms, including their own, and inspire others to succeed and pursue organic farming. "Most of our customers are looking forward to experiencing the simplicity of farm life," Megan says. "We encourage them to pick vegetables and fruits from the gardens for their meals. They enjoy hiking in the hills behind the farm, and playing a game of horseshoes. Staying with us in our wall tents is a rustic adventure and that is their biggest appeal. Our farm is a working farm and looks accordingly. Our offices are dispersed between a mobile home and the barn. We just add a few flower bouquets, special touches, and the farm is what brings our customers."

Challenges are always part of such an adventure. One of them, Megan says, is for the owners to find time with the added B&B element to enjoy the simplicity of farm life themselves. "It takes time and energy in order to do the kind of job in which you'll

come from you and your own legal advisors, customized to your own specific needs once they've become clear.

Selling value-added products

A farm B&B lends itself well to farm-crafted, retail products. You may not have planned for this at all, but may find it to be true and choose to re-arrange some of your current farm produce or skills into B&B-friendly value-added products as guests begin to request and comment on certain items they see or use during their stay. These could include:

- Gift baskets of farm-made non-perishable foods, bath products, or candles;
- Furnish and decorate B&B rooms with products that you sell retail with an order form in the room. These could be your own cross-stitching or a gallery of art, blankets or quilts on commission from local artists.
- Your farm's own cookbook;
- Hand bound blank journals with covers decorated with your farm's pressed flowers

No B&B plan fits all, but there is common ground.

A farm B&B in France walled off the parking area from the rest of the farm, and parks these wagons there for newly arriving guests to load up their luggage and take it inside the farmyard that leads to their room.

CHAPTER FOURTEEN

Chapter Resources

Contacts:

MaryJanesFarm
1000 Wild Iris Lane
Moscow, Idaho 83843
   www.MaryJanesFarm.com

Eckert AgriMarketing
Jane Eckert
8054 Teasdale Avenue
St. Louis, Mo 63130
   www.eckertagrimarketing.com

Better Business Bureau
Free online articles and fact sheets
for USA small businesses, including how to create custom business
plans with samples
   www.sba.gov/library/pubs.html

---

have return customers, as that is the goal," she says. "However, it is also very rewarding. After each of our guests leave, we repeatedly are saying how wonderful it is to meet each of them and how inspiring they are."

While B&B types are many, those who love the business seem to have one thing in common. They love their farms, and they love the rewards of working with people. "This last summer we had a mother come with one of her eight-year-old twin daughters," Megan says. "The following week she came with her other eight-year-old twin. It was important for her to have one-on-one time with each of her daughters to experience the farm. They enjoyed the swing, visited with all the animals and picked fresh vegetables. One of our two-year-old visitors had a choice between homemade cookies and picking her own tomatoes. She promptly headed for the tomato patch and ate all that she picked. For a farmer, there isn't anything more rewarding than that!" Sometimes, the support resulting for the farming population generated by successful farm B&Bs is immeasurable. ❧

Hosting Professional Chefs 15

"When chefs see the way we do things," says farm owner, Riley Starks, about producers giving tours of their operations to chefs, "they see the superiority first hand, and because of that, we have a customer for life."

And along with witnessing the superiority of the product, chefs get a close-up of the "farming story." As discussed in Chapter 10 about the benefits of sharing the farm's story through a journal as a form of marketing, this story-sharing is also of great importance for farmers producing for chefs. According to the Leopold Center for Sustainable Agriculture, new emerging markets are going to drive a new farming ethic towards greater chef/farm connection where this story becomes very important. Their Leopold Letter states, "The Hartmann Group reports that 62 percent of the consuming public now wants to buy food that is 'consistent with their values,' and leading chefs throughout the United States are telling us that success in the restaurant business is now 'all about the story.' "

Farmers who supply chefs as at least part of their customer base have found that hosting chefs directly on the farm in a variety of ways boosts business dramatically. Riley Starks, owner along with his wife, Judy, of Nettles Farm gives tours to chefs of their eco-farm and other unique, environmentally-friendly food-harvesting methods wildcrafted from the land and sea. Nettles Farm operates on very small acreage on Lummi Island, one of many islands off

the Northwest corner of Washington State. From this relatively isolated island that must be accessed by small ferry or private boat, the Starks supply chefs across the country, as well as local islanders at the Saturday farmers' market, and three of their own, on-island restaurants: Willows Inn, The Beach Store Cafe, and the Taproot Pub and Espresso Bar. Their produce includes heirloom tomatoes, washed salad greens, and assorted vegetables including asparagus.

After farming and delivering to restaurants for a number of years, the Nettles Farm owners added a certified kitchen to handcraft on-farm pasta, ravioli, and lasagna from their farm's ingredients. The recipes for the ravioli and lasagna change with the seasons, starting with wild-crafted nettle in early spring, and moving on to include other herbs and vegetables as the growing season progresses. Chefs take pride in the foods they provide their customers, and to see the beautiful gardens of Nettles Farm—the tomatoes thriving in the greenhouse, the nettles hanging to dry, and the certified on-farm kitchen as one hears the history direct from the farmer—one can't help wanting to be part of this story.

Dylan Stockman, a professional cook, has spent time touring sustainable farms, including a certified organic apple orchard, and Quillisascut Cheese Company, a small family eco-farm in Northeastern Washington State. Quillisascut's motto reads, "Cheese from the Pampered Pets of Pleasant Valley." Owned by Rick and Lora Lea Misterly, Quillisascut specializes in handmade, artesan goat-milk cheese. Dylan describes the power of "the story" after his visit there. "I believe that cooks and chefs have so much to learn from opportunities like the farm tour," he says. "They are able to see where their product comes from. They can appreciate the life of the tomato, beet, or other products. When you harvest your own, you have a sense of pride, because you harvested it and you get excited about what you are going to do with it. You don't start to think about 'gourmet;' you just want to do it justice; you want the product to stand out, so people can say 'Wow, I have never tasted a pepper so rich and sweet, where did you get these?'—and then you are able to tell them. Then the customer guests get excited because they can read your excitement."

Working directly with chefs

The relationship between farmer and chef is unique. Adam Childs is Nettles' main farmer, living on the property during growing season. He interacts with the chefs he grows for locally, as well as the ones who come from further distances to tour the farm, such as group chef tours arranged by the Chefs Collaborative (see resources). I asked Adam about the differences between farming for chefs, allowing them to come to the farm, and other types of market outlets, such as farmers' markets.

"The nice thing about my relationship with Chef Craig Miller," Adam says about the chef he works with most regularly, "is that we're friends as well as co-workers. He doesn't give me formal requests for things; it's more that we have loose conversations about different ideas. For example, this year he suggested that we try growing peas just for their greens (for a garnish), mache, and Jerusalem artichoke. So far the pea greens and mache have been successful. I haven't figured out whether Jerusalem artichoke is going to be a viable option. Often these loose conversations happen when he comes up to the farm once a week. We walk around the farm and talk about what will be ready for the weekend, and what will be coming up the next week."

Adam makes many suggestions to the chef: "Because I'm the one who looks through the seed catalogs and orders the seeds," he says, "I'm constantly seeing vegetables that I either haven't grown or are newly offered. It might be closely related to something we already grow, or it might be something that I just think would look really nice on a plate. The majority of the time, Craig is open to trying anything at least once, especially if it's something rare or unusual."

Making strong chef connections can ease the job of selling. "The produce I grow for the restaurant," Adam says, "I know will be purchased and used. (However) the things I grow for the market… growing it is only half the job. I still have to sell it. Also, when I go to the Saturday market, I sell more conventional vegetables like lettuce, peas and beans. Working with a chef like Craig and the types of food they cook at the Willows, I get to grow things like escarole, frisee, hon tsai tai, tat soi, and broccoli raab. Craig uses these types of vegetables regularly, which at a market only might interest one or two customers."

Riley and Judy Starks progressed from farming and distributing to other restaurants, to wanting their own restaurant on their own property. However, that proved impossible because they share a driveway with neighboring acreage owners who didn't like the idea of the public driving up their private path. So when the Willows Inn, which was a slight distance from the farm, came up for sale, the Starks purchased it. Other farmers who don't share ownership in the restaurants they grow for, or don't have quite as tight of a relationship with their chef customers, suggest spreading the customer base to several chefs. If you grow for a particular chef or restaurant owned by another, and that chef moves or the restaurant goes under, you're left with perishable produce aimed for a niche market, scrambling for customers. An ongoing on-farm connection with several chefs can help safeguard from this situation.

The chefs have much to gain by working with you, including that great story of earth connection, history and sustainability to share with customers. "I think that what makes what we're doing unique," Adam says about his work on Nettles Farm, "is that the food that goes to the restaurant is never packaged; it travels less than a mile, and all the compost from the restaurant comes back to the farm, and all the chicken manure from the farm is used on the farm. It's a very small loop." By visiting farms, chefs can discover the interrelations needed on sustainable farms and come to appreciate a farm's eco-system and the vast knowledge this type of farming takes.

As an example, Adam says that keeping chickens on the farm just for compost is well worth the expense, even if they're not laying or being eaten. Riley points out to visitors how the chickens saved their tomato crop. Each year, Nettles tomatoes had been grown in the ground under the same greenhouse. "One year," Riley says, "We had a blight, so instead we put chickens in the greenhouse with wood chips." Tomatoes were grown in a different location that year. "That completely eliminated the blight," Riley says. "The carbon from the woodchips mixed with the chicken manure seemed to do the trick by completely changing the structure of the soil." Today, the tomatoes grow strong and healthy in that original greenhouse.

When chefs hear stories like this, they learn they can trust their farmer to understand the complex science behind providing them with sustainable produce, and can see why their produce is well worth paying good prices for. Once again, they can pass this story on to their restaurant patrons. This "passing of the story" is international in appeal. While visiting French farms and their associated restaurants with my husband, the chefs themselves came out to the guests to describe the origin of their menu's ingredients. Their descriptions of ancient cow-milking breeds, walnut groves and Normandy's apple cider history were fascinating.

Hosting chefs on your own farm

When hosting chefs at an on-farm tour, show the foods they prepare (or the ones you hope they'll

Riley Starks, owner of Nettles Farm, shows his greenhouse tomatoes to visitors.

CHAPTER FIFTEEN

eventually purchase from you) actually growing, and if applicable, demonstrate how they're dug up or picked and prepared for restaurants. Explain how you keep your soil alive to keep the produce healthy and delicious, and what you do in emergencies, such as unexpected droughts. Tell your farm's history, the history of the crop, especially if it's an heirloom, and why it's a better choice for culinary reasons. Find some hands-on activity the chefs can participate in, whether simple taste sampling or harvesting baby carrots.

At Edith Walden's Willowrose Bay Quince Orchard on Guemes Island—another island off Washington State—Edith led visitors through one of the few tree-ripened quince orchards in the country. At the end of the tour, she allowed guests to harvest another unusual fruit, aronia, from a patch that was experimental that year and wasn't being harvested for production yet. Plus, she brought out Quince and aronia syrups and preserves to taste on crackers or to pour into sparkling iced water, and sent visitors home with recipes.

Also, understand the pressures and uniqueness of the chef's business as well. On my own Island Meadow Farm, a particularly artistic chef wanted red and yellow pear tomatoes. Not the orange ones I had in surplus, nor the popular heirlooms—red and yellow pear shapes, period. He was envisioning a specific way he wanted to use them on his plates. Chefs sometimes treat the plates they fill for guests as art canvases, with their guests expecting a good show for their money. It's valuable to them if they know the farmer appreciates this and wants to help.

If you've never attended a farmer's tour, it may help you to attend one before you put one on yourself. Check local "happenings" in your newspaper, call your local agriculture extension for farm tours, or contact Slow Food (www.slowfoodusa.org USA, www.slowfood.com international) to see if there is a "convivium" near you that's organizing a tour. Notice what you, as a visitor, enjoy about the farmer's presentation, and what you would have improved if it were you; then adapt your own plan.

On-farm chef schools

When Dylan Stockman spent time at Quillisascut Cheese Company, he was participating in their on-farm cooking school. An on-farm culinary school or workshop can bring in substantial revenue as a complementary home business even beyond gaining faithful, future chef customers, as all the while it helps preserve local, sustainable farms. "The thing that most chefs and cooks don't know," Dylan says, "is that when you buy from the big, monstrous produce purveyors, you are giving them more money and power to push the little guys out. Those little farms then have to sell the product to the big guys." As many small farms know, direct selling specialty products versus wholesaling mass-produced crops helps keep them in business. "Then there is the problem with maybe even having to sell the farm," Dylan says, "because they can't move their product. It is very scary and it is something I am always keeping my eye on. We all need to work together, to support our local economy. We are all responsible."

Setting up a cooking school or workshop for serious, professional chefs will take research beyond the scope of farming itself, of course. To teach cooking, your local area may require a food handler's permit (usually easily and inexpensively obtainable, it's a good idea for all farmers involved in food crops to have this permit) plus a certified kitchen, and perhaps other regulations with which to comply. Check with local licensing and permit authorities, as well as your local agricultural extension. Quillisascut teams up with a talented local chef for their farm's cooking school segment. The chef comes to their farm, but for other farms not ready for this, there is also the possibility of starting out with a chef providing a nearby certified location for cooking classes that include first visiting your farm to harvest the ingredients. See Chapter 13 for more information on emphasizing education on your farm, and the Epilogue that discusses Chef Sally McArthur and her connection to farms in the USA and France.

Remember the professional personal chefs: Not all professional cooks and chefs work for restaurants. In this age of entrepreneurship, there appears to be an increasing demand for "personal chefs" who cook for families or private home gatherings in their customers' own homes, sometimes specializing in at least a few of their own kitchen creations. The same market driving restaurant chefs to purchase from sustainable, local farms with good stories to tell could infiltrate this arena as well.

Finding chefs

To reach personal chefs in the USA, contact the United States Personal Chef Association at:

www.uspca.com.

For restaurant chefs interested in local farm produce and chef tours, national chain, franchise restaurants may not be the place to start looking. Their cooks are usually following a corporate-stipulated formula, and are not allowed to purchase locally, nor do they have artistic freedom to choose their own ingredients and design their own menus (there are exceptions, of course). Also, cafes that specialize in low-cost foods purchased from cheap generic wholesalers most likely wouldn't be your choice for finding chefs.

For local chefs, seek nearby, higher end restaurants and ask the names of their chefs, especially their weekend chef. You can drop off fliers and free samples to your list of chefs, and offer to host local chefs casually on a request basis according to whenever their schedules work with yours. Often, the most important chefs work the weekends, and begin preparing early Friday morning. This time has proven the best for many farmers who sell to chefs as far as dropping off an initial free sample and literature describing the farm.

Also, professional culinary schools, even a slight distance from your farm, may provide potential chef tourists if you can present them with detailed and professional literature on what your farm offers during the tour.

For more distant restaurant chefs from the USA, or more formal tours for groups organized by established entities, consider membership in The Chefs Collaborative:

www.chefscollaborative.org

This is a national program that can help bring farms and chefs together. Riley Starks stated that he connects with chefs both independently and through the Chefs Collaborative, of which he is a member.

One of the Chef Collaborative's chapters in Oregon, USA has joined an organization called Ecotrust along with several other regional sustainable organizations. This network formed the Farmer-Chef Connection, which is a great regional model that any region can learn from. It finds numerous ways, including on-farm chef tours, to connect chefs with sustainable farmers. The Farmer-Chef Connection can help those in other states set up a similar program.

"Ecotrust and the Portland (Oregon) chapter of the Chefs Collaborative facilitate the Farmer-Chef Connection," says Debra Sohm, Director of Food and Farm Market Connections, Ecotrust. "It is a replicable model that can be implemented nationally, and consists of a workshop, survey, and direct marketing guide. The program promotes experiential learning and gives producers an opportunity to interact with chefs and retailers. Producers and buyers learn how to sell/buy to and from each other. The program provides buyers and sellers the following: a chance to network; learn about issues (communication etiquette, product handling techniques, etc); meet new potential business partners; and hopefully reach agreements that can lead to the beginnings of a relationship."

Each connection strengthens the whole

Debra Sohm illustrates how profound even a single farmer-chef connection can be when she makes the following statement about food producers establishing strong relationships with their purchasers, "The collective strength and viability of the regional food economy grows with each new individual relationship that is formed."

Dylan Stockman knows first-hand what an impact the farmer-chef connection can have: "I like to look down at my dirt-caked hands, and say, 'Yeah, that's where it's at…'" Dylan says about working in the garden. "You know you have accomplished something; your basket is full along with your heart. If all the cooks in the world could get a sense of that feeling just once in their lives, I honestly believe the kitchen would be a better place."

Opening the doors to selling to restaurants

Selling directly to restaurants, according to many farmers who do this, represents the state of the art in growing: an incredible feeling to grow great produce and know you're getting it to where it's really appre-ciated. For chefs, cooking from the garden represents the ultimate in cooking. According to Alice Waters, owner of the Chez Panisse restaurant in Berkeley, California, buying direct from farmers had a dramatic impact on restaurant operations. The restaurant chefs have become fully committed to cooking seasonally and planning the menus around what's fresh. "We cook from the garden," Alice says.

When you visit restaurants, bring samples and a cutting board with you. Prepare your product for the chefs and let them taste your great produce. Especially with a new product that requires special preparation, make sure that you train the chefs in how to prepare it; otherwise they may prepare it incorrectly and not order it again! Stress the five-day-a-week delivery service (if this is what you plan to offer); the special varieties; the freshness, uniqueness, and vine-ripened flavor of your products; or the extra care and personal attention your products receive. Order extra copies of seed catalogs to leave with chefs.

Products must be of top quality and freshness, and available as needed, sometimes in small quantities. Other restaurant concerns are for price, consis-

Farmer Adam, with Nettles Farm, explains his organic farming and careful harvesting techniques to visitors.

Contacts and Resources

Nettles Farm
www.nettlesfarm.com

Quillisascut Cheese Company
www.quillisascutcheese.com

New Farm online article
"Farmer insight: Building chef connections"

www.newfarm.org/features/0103/california/beard/index.shtml

www.chefscollaborative.org

www.slowfood.com

tency and reliability of supply and delivery, and for specialty produce not available in wholesale markets. Be prompt with deliveries and supply exactly what you've promised. Reliability is a must. If you don't show, it means panic in the kitchen and missing items on the menu. They may not buy from you again!

One of the most enjoyable aspects of selling direct to restaurants is the ongoing dialogue with food professionals. Feedback from chefs allows the farmer to upgrade his products and services. Always communicate if there is a problem. If you are short on a product, call the restaurant to let them know so they can cover themselves on the menu. Chefs often are unfamiliar with the new specialty crops coming into the marketplace and are usually grateful for recipes, cooking suggestions and nutritional, storage and handling information.

Winter is a good time to sit down with chefs, show them seed catalogs, and plan varieties and quantities. Regular review meetings allow you to discuss what you grew that was right, and what you grew too much or too little of. In addition to quantity and variety, the meetings can decide how things are grown, picked and packaged. Salad greens for example, can be plucked either as small, tender edibles, or allowed to mature for greater color, tone and texture. Size is important: green garlic, for example: how many big ones to be chopped up for cooking, and how many small ones to be used as whole vegetables? Some chefs are still accustomed to buying from brokers, and it will take them awhile to understand the lead times you need in order to grow crops, which is another good reason to find ways to host them on the farm. Give them a harvest calendar of upcoming crops so they can plan menus and purchase from wholesalers the items you can't supply.

Taking chefs on a tour of your farm (as this chapter illustrates) is a good way to impress upon chefs the farm-freshness of your products, to give them a sense of how things are grown, or why you can't get certain products to them at certain times of the year. The ultimate goal of the grower-chef relationship is the realization of mutual dependency and widening the circle to include consumer education about the importance of local agriculture and green belts around cities. When this trust is established, the chefs will also realize the importance of the agritourism outreach you have established at your farm. ❧

– "Opening the Doors to Selling to Restaurants" is adapted and updated with permission from Selling to Restaurants, in *Sell What You Sow! The Grower's Guide to Successful Produce Marketing,* © Eric L. Gibson.

The Dairy: *From cheese-making classes to school tours* 16

The author knew she had found the treasure she'd been seeking in searching for an exceptional dairy involved in agritourism. It had to be a replicable/adaptable model from which others could learn, and the farm itself successfully sustainable. But even more, it had to fit her definition of authentic agritourism: the farming itself is center stage; agritourism is a compliment. Then I came across a spectacular, grass-fed eco-dairy that successfully used a rotational grazing system, and also added agritourism. Did their farming remain center stage? Digging deeper, I found the following notation on the website of Sweet Grass Dairy in Thomasville, Georgia, USA.

> "We are a working farm. When coming, please do not wear heels and dresses. You might step in something that will not compliment your outfit."

Bingo!

Sweet Grass Dairy, owned by Jessica and Jeremy Little, is a 140-acre goat dairy and handcrafted cheese facility where the goats graze outdoors on pasture year-round. Its sister dairy, Green Hill Dairy (about 30 miles away), is owned by Jessica's parents, Al and Desiree Wehner. Green Hill Dairy is 340 acres and also allows its Jersey cows to graze outdoors year-round with rotational grazing. The milk from both the cows and goats is crafted into cheese on Sweet Grass Dairy, and this is also where, for the most part, the farms' agritourism is centered.

Sweet Grass Dairy holds two yearly open houses, gives farm tours by appointment, and offers cheese-making classes.

A brief history of the farms

Sweet Grass Dairy was actually started in the year 2000 by Jessica's parents, Al and Desiree, long-time dairy people and then also owners of Green Hill Dairy. "My father is from a long line of German farmers from Western New York," says Jessica. "He moved to Athens, Georgia to attend the University of Georgia since he thought he wanted to be a vet. My mother is from Gainesville, Georgia, and after they met, they both ended up with degrees in Animal Science."

Her parents decided they'd rather be dairy farmers than vets, a choice that would eventually make an incredible contribution to sustainable farming. "After they both graduated," Jessica continued, "they had the opportunity to run a dairy farm in North Florida–conventional style, free stall barns, milking 1,200 Holsteins three times per day, growing and harvesting all of their own silage, and basically, working themselves to death."

Eventually, Jessica says the owner let her parents buy into the farm as partners in 1986. "After a few years, my parents thought that there had to be a better way—the animals were not as happy as they should be, the employees were not as happy as they could be, and the soil was not as nutritious as it

> *When my mother saw the quality of the milk from cows out on grass versus cows in barns—as most universities teach as "the best way" to farm—she felt the need to show other people. She wanted to show people what makes food good and why it is important to know where your food comes from.*
>
> *– Jessica Little, Sweet Grass Dairy*

should have been. After hearing a guest speaker in Wisconsin on rotational grazing, my dad returned to the farm and began telling my mother how much the idea made sense and how he wanted to get out of the partnership and start a New Zealand rotational-style dairy farm."

In 1993, Jessica's parents bought land in Brooks County Georgia and started Green Hill Dairy. "The 2,000 Holsteins were replaced with about 500 Jerseys. The 340 acres were divided into five-acre paddocks where the cows were rotated every 12 hours."

"Most dairy farmers in the area believed they would be out of business within a year or two, but my parents believed in the fact that cows should live out on green pastures, graze lush grass, and not have the stress of confinement farms… Green Hill Dairy has proven to be a great success."

After deciding to add artisan cheese to their farm's products, and then falling in love with goat cheese, Jessica's parents started Sweet Grass Dairy, where goats were grazed and cheese was crafted. Soon, Jessica and Jeremy accepted their parents' invitation to operate Sweet Grass Dairy, and enjoyed it so much; they purchased it from their parents.

The beginnings of agritourism

It seems the agritourism segment began with Desiree's desire to let others know the value of sustainable farming. "My dad," Jessica says, "always (states that) the first important thing he had to do,

after purchasing the land to create the dairy, was to get the soil nourished again—the farm was an old cotton and peanut farm before and the minerals were so out of balance. He always said that farming starts with the soil." Her mother felt that others should see the high quality that comes from such sustainable farms. "When my mother saw the quality of the milk from cows out on grass versus cows in barns—as most universities teach as "the best way" to farm—she felt the need to show other people. She wanted to show people what makes food good and why it is important to know where your food comes from."

The open houses

"We have two open houses each year—one in the spring and one in the fall—called Market Day at Sweet Grass Dairy," Jessica says. "We invite other local producers that follow our same farming philosophy such as White Oak Pastures, Turkey Hill Produce, Zebra Truck Farm, Magnolia Hill Soap Company, O'Toole's Herb Farm, Tiger Mountain Vineyards, and Broken Arrow Honey Farm, to name a few. Market Day is free to the public and we don't charge a vendor fee for the people who are selling their products. We give tours all day long as well as have an animal petting area with cows, goats, chicks, sheep, pigs, and ducks. We also offer samples of every single one of our cheeses."

Jessica says these Market Days are very profitable for the farm, even though they are free to the public, and that people from quite far away come to the event. "Our Spring Market Day is a little slower since it is around the end of April or beginning of May, and people are really busy, whereas, our fall Market Day is crazy since it is the Saturday after Thanksgiving. It's a lot of work to get ready for but it is worth it."

Tours by appointment

"The other part of our agritourism is our tours of the farm and cheesemaking facility," Jessica says. "We

Visitors can view the cheesemaking at Sweet Grass Dairy through a glass window.

The tours have proven very popular. It appears there's more demand than supply at this point. "This part of our business really has taken off and if we wanted to do tours every day for those months, I'm pretty sure that we could with a little advertising. I think that there are tremendous opportunities in this (agritourism) area of the business," Jessica says.

Cheesemaking classes

Jessica's husband, Jeremy, is in charge of the cheesemaking classes, which are a great example of allowing the agritourism segment of your farm to be made up of your other hobbies or passions that synergize with the farm. "Jeremy has a degree from Florida State University in psychology," Jessica says, "but ever since I've known him, he has wanted to go to culinary school to be a chef. He is an amazing cook and really loves food. We met while working in the same restaurant in Tallahassee when we were both in college in 1999." Artisan cheesemaking has helped give Jeremy a culinary outlet.

Jeremy puts on his cheesemaking class about every month or so. "We only take about 15 people and the class is on a Saturday from about 8:30 a.m. to 4:00 p.m. Depending on what he needs to make and what the focus of the class is, Jeremy will select two different cheeses to make. People love it and the majority of the attendees come from Atlanta, Georgia. We charge people $75, but it includes lunch, all the supplies needed in the cheese room (aprons, boot covers, hair nets, etc), a cheesemaking book, and we send them one of the cheeses that they made when they become legal—at least 60 days. (At this writing in the USA, cheese made from raw milk must be aged at least 60 days.) It's a fun day as well."

are pretty much booked for Thursdays, Fridays, and Saturdays for tours from February though April. We do school groups, church groups, retiree groups, and traveling groups. The tours normally last one-and-a-half to two hours, with the group being able to see the milking of the goats, cheesemaking, and taste the cheeses, and then spend some time out in the field with the animals."

A special component of Sweet Grass Dairy's farm tours is that the fascinating science and art behind their sustainable dairy method is revealed. "We hope to provide a glimpse of the whole cycle from the soil to the cheese," Jessica says.

Sweet Grass Dairy charges $5 per person over age three, and requires a minimum of 20 before booking an appointment. "We get a lot of repeat tours from the same teachers each year," Jessica says. "It's a lot of fun to teach these kids about where their food comes from, and most have a ton of questions regarding the animals and process. We also do a limited amount of tours in the summer for summer camps, but I try to steer away from scheduling too many, due to the heat."

Marketing the agritourism

To generate farm visitors, Jessica says that their local Chamber of Commerce has been very helpful. "For our last Market Day they were able to get a little segment on the local morning news in Thomasville and Tallahassee. They are able to help us advertise since we do not have an advertising budget, and they get the write-ups into the correct person's hands. I really enjoy working with them."

The farm's own literature further adds to the marketing campaign. "We also use our website to advertise and get the word out about Market Day and group tours. Our local printers do up a poster for us and we plaster them around town for Market Day."

Jessica explains that networking with others also has proven successful. "We do a few other events where we join other people's open houses, and they usually do the advertising for those."

As far as keeping customers interested once they've discovered the farm, it's also been valuable to continue to communicate with visitors. Sweet Grass Dairy uses newsletters and the collection of customer feedback to keep their agritourism business progressing. "We have a newsletter sign up sheet for every event that we do," Jessica says. "So for example, after Market Day, we might see that we had 30 people sign up for cheesemaking class infor-

Goats and family dog rest in the shade on Sweet Grass Dairy.

mation. Out of those 30 we'll have 15 or so that will be able to make it to the next cheesemaking class."

Business planning, safety, and liability

Sweet Grass Dairy's agritourism is one of those that evolved on its own without a previous business plan. "The demand for tours really created this addition," Jessica says. "Our first Market Day was in July of 2002 after Jeremy and I came (to Sweet Grass Dairy) because business was really slow and we needed a way to make some money to cover our note payment. Cheesemaking classes started with so many people asking if we would do one, and the first was in 2003. Our tours for school groups were also started out of the phone constantly ringing with requests to bring out their students; home school groups, or church groups."

However, it's never too late to develop such a plan. Many feel it helps them focus and expand, even if their operation is relatively small. "I would like to start working on a formal business plan to figure out how to make it better for people," Jessica says.

The local Health Department oversees safety standards for the farm. No one is allowed into the commercial cheesemaking facility, but, rather, they view it from the outside looking through its windows. "We have a great insurance policy and we have people sign sheets stating that they have read our sanitation program and also that they will follow it for the cheesemaking class." Jessica says they are also considering other forms of waivers. As rules, regulations and society itself changes, it's always good for farms to keep up with business and liability policies that work best for their particular farm and location.

The agritourism payoff

"I think that the long term benefits are really endless when it comes to agritourism," Jessica says, "and there will be more of a demand in the future." Jessica and Jeremy would like to continue to expand

An artful display of farm-made cheese is attractive to farm visitors, this photo is used on the poster to announce Market Day, an agritourism event at Sweet Grass Dairy.

and improve their agritourism segment, including a camp for children. They share concerns with those who want to restore humankind's connection to real farms, and help eliminate stereotypes and the deterioration of the general population's knowledge of authentic farming. "I have been to some 'farm' tourism spots where the animals were not cared for very well; it was dirty, and I just shook my head in sadness to see that the people that paid to go to the farm really believed that this was 'farm' life. It is a fine line in charging people for your time while you give an educational tour on a farm versus charging money to provide entertainment on a farm. We lean on providing educational experiences instead of providing entertainment to meet instant gratification."

Although many agritourism benefits for the farmer can be very long term, there are quicker benefits as well. In Sweet Grass Dairy's case these include generating farm-class students and the direct sale of products. "Market Day is free," Jessica says, "but we sell all of our products to people and these have proven to be very profitable."

When asked about favorite memories concerning agritourism, Jessica says there are so many, and then has to consider which one to describe here. "Last year we had a really cute group of Girl Scouts come out for a tour one day not long after we started kidding out," Jessica says. "We had about 20 baby goats all in need of names, and they proceeded to name them all with a food related theme. They were really cute and I still chuckle every time I see Snickers and Oreo in the field. Also, I love seeing the children so happy when they play with the animals. The way that they beg their parents to take home a baby goat is priceless. The goats are just as happy!" ❧

Resource

Sweet Grass Dairy
  www.sweetgrassdairy.com

Beyond wagon rides: 17
Healing arts on the farm, and a fresh look at horses and petting zoos

When is a professional massage more authentic to a farm's agritourism package than a hay ride? Perhaps when one of the farm owners is a massage therapist (or wants to be), and the farm doesn't grow or use hay.

This chapter explores an emerging idea of adding various healing arts (bodywork, day spas, healing and spiritual retreats, the healing interaction with animals, including horses) to the farm's agritourism package. It also furthers the idea that the farm is authentic according to the farmers themselves, not the stereotypes of what a farm should be.

Your farm and its location are the major attraction and quite often, nothing else is needed. But "value added additions" to the agritourism venture could draw on the farm owners' other passions, career dreams, hobbies, and even extended family members who want to join in on the business, rather than adding generalities such as hay rides if your farm never uses or grows hay, and never plans to. "Real" hay rides come from harvesting hay and allowing weary farm workers to ride on the hay-loaded wagon on the way to the hay storage barn. Or, they come from loading up a wagon with hay to feed the ranch animals, with the farm's children riding atop the hay on the wagon to push hay into the fields.

As with the dairy farmer in the last chapter who had a second dream of going to culinary school, and then came to love the cheesemaking classes he taught on the farm, it isn't always necessary to add

other "farm" entertainment if it becomes just an added burden and brings no real pleasure to the farm owners.

As a part-time bodycare practitioner, nothing was more pleasant for this author than getting to stay home on the farm to give healing sessions, along with the added bonus of generating word-of-mouth promotion about our farm. People who otherwise wouldn't have known about it, or even known to look for a local farm, came across it this way and were pleasantly surprised. Some farms, as you'll read below, provide beautiful settings for healing centers or day spas. The setting itself is a healer. Almost immediately, it can take customers away from their harsh realities even before their professional healing sessions begin.

Several years ago, Mary Anne K. McIntyre, Albert P. Sabatini, and Kevin and Sara Ruch, came across Pennsylvania farmland that had been owned by the same family for more than 100 years and had been a working farm up until about 30 years ago. Thanks to this foursome who bought it and began cover cropping the land to build up the soil, it is a working farm again called 14 Acre Farm. This time around, the new farm includes White Bear Bed and Breakfast, among other value-added services. Mary Anne is a professional physical therapist, but is also trained to give massage, and one of the farm's B&B rooms has been turned into a massage room where she can enjoy offering her skills as a massage therapist. "It occurred to me to offer massage for B&B customers because what better way to assist people to relax and feel that they are being pampered?" Mary Anne says. "We all live crazy busy lives and massage is a great way for people to truly relax." She's able to give guests massage at a good price, and they don't have to travel off the farm's B&B to receive it. Massage is listed on their B&B web site as an amenity. "If the guest is interested," Mary Anne says, "they inquire about the service. Then they are given more information and price, and they can schedule anytime during their stay. Most of the guests seem to prefer Saturdays."

Day spa on a lavender farm

A trip to Anacortes, Washington State takes ferry boatloads of tourists to a group of evergreen islands which are accessible only by the ferry, private boat or plane. On one of these spots of paradise named San Juan Island, the summer explodes with the heavenly scent of lavender grown on Pelindaba Lavender Farm—Pelindaba meaning "place of great gatherings" in South African Zulu. The farm's owners have succeeded at creating a destination farm complete with a lavender festival, lavender gift shop and on-farm activities for guests. Visitors, from children to elders, enjoy walking amidst blooming lavender sprays, picking their own varieties to take home, observing lavender cooking demonstrations, and participating in herbal craft-making. The farmers extract the essential oils of their lavender in their

own on-farm, custom-made distillery. From the lavender and its essential oils, they handcraft products for culinary, personal, therapeutic, and household purposes, and even pet care.

For a couple of years, a day spa was remarkably successful on the farm. The owner of the day spa shares how this mutually beneficial partnership helped the farm owner receive rental revenue along with a strong customer attraction that they didn't have to run themselves, and could therefore concentrate on farming. But the idea of a day spa or other healing center on a farm with a healing environment can work whether the center is owned and operated by the farmers themselves, is a simple and occasional offering by a farm owner such as Mary Anne's described above, or the business is separately owned by another and the farm is just rented as the location. Any of these especially work well when the farm's crops, such as lavender, can be used as part of the healing center's products.

It was Pelindaba Lavender Farm's beautiful purple fields and views that inspired massage therapist, Ciely Gray, to envision an on-farm day spa. "The

A rustic, yet attractive, sign works well to lead visitors to a farm's healing center.

farm was the inspiration for the spa," Ciely says. "I had been doing chair massage at the farm in the summer and saw what a beautiful destination it was. I approached the owner of the farm about the idea, having no idea where this massage venue could be located on the farm property. He owned a private home on the property which was being rented at the time. A few months later the renter decided to leave, and the home became available. I then approached some key therapists on the island to help develop the concepts and services. Thus, the spa is therapist owned & operated."

The spa was not owned by Pelindaba Lavender Farm per se, but in collaboration with it to evolve both businesses by sharing advertising. Ciely found the house to be exceptionally welcoming to visitors of the day spa. Initially, it needed new paint, some new electrical outlets, and of course, some decorating. But even furniture, in this case, was provided by the farm owners. "It was amazing how beautifully the house was set up to create a spa," Ciely says. "When clients walk in the door, they walk into a cozy living room. It's very humble in that sense. They are then offered citrus water or hot tea in the winter while they fill out a health history form for us, and take a minute to relax before their treatment."

While a farm location came upon Ciely as the initial idea for a day spa, this idea works when owners of off-farm healing centers or day spas already in existence are looking for rural or natural places to expand or relocate. If they already have a good reputation in another location, they can bring their reputation as a new "crop" to your farm, thus drawing more customers to the farm. In Ciely's case, she took advantage of being brand new with no name yet, allowing the farm's customer-pleasing crops to actually inspire the day spa's name: "We wanted something unique, pertinent to lavender, and simple." Ciely eventually chose the name, "Lavendera." "I found out after we came up with the name that it means washer woman in Spanish. I

A lavender farm and day spa can make good companions. Nature enhances the relaxing setting of the spa. Visitors are led to the fields and the day spa, giving them more than one reason to enjoy visiting the farm.

liked the humble reference this implied. I also liked the international flavor and sound of Lavendera."

Sharing the promotion with the farm and other businesses

When asked about the benefits of overlapping the marketing with the farm, Ciely responds, "It can boost both businesses, save money with shared advertising, and has double the draw." Pelindaba Farm gave Lavendera its first exposure, and eventually the farm and day spa connected with other local businesses to create packages and share advertising. "We have worked with the B&B's and the motels on the island," Ciely says, "creating packages such as a 'Couples Getaway,' that include nights at their establishments, possibly a kayak trip, and a voucher to be used at the spa. We share links on our web sites if we have created a package together." Next, Ciely says she is looking into collaborating with the local bike tours, Elderhost tours, and more kayak tours.

Bill Varney, owner with his wife, Sylvia, of Fredericksburg Herb Farm in Texas, grows organic herbs, flowers and vegetables, and also offers an on-farm restaurant, bed and breakfast, and a day spa called Quiet Haus Day Spa that offers aromatherapy with

Swedish massage, body wraps, and European facials. For these farm owners, Bill says he prefers to own the day spa business himself, rather than rent such a business out to another (see Chapter 5 on choosing a business entity and check with your accountant and attorney to see if in this case, the farm and healing center should be separate businesses that you own, or if they should both be under one business entity's roof).

Whether the day spa business is owned by you, the farmer, or another, Bill agrees that an on-farm day spa can be a win-win situation for both businesses, each enhancing the other. "It is a more laid back feeling," he says about the experience customers have when coming to an on-farm day spa, "and many people are inspired to walk the gardens, relax, and feel closer to nature." It's no wonder that visitors to Quiet Haus Day Spa are inspired: Among the farm's charming old German garden ambiance, they'll find clusters of buildings bustling with farm and garden activity, such as the Poet Haus, where candles are handmade from the farm's herbs, plant extracts, oils, and beeswax. See their website at:

www.fredericksburgherbfarm.com.

Is your farm the right setting for a healing center or day spa? Smaller and eco-friendly farms are great settings for drawing crowds interested in health and escaping the adrenal-overdrive of modern living. "I would say to be sure that the farm is easy to locate and get to for customers," Bill Varney says. The customers that come to farms that offer healing and retreat may be different than those seeking adventuresome tours of active working ranches or the chance live in rustic surroundings in nature. But in some cases, both adventure and retreat can be offered and make a wonderful match, such as with a ranch that offers rose scented hot tub soaks and reflexology sessions at the end of a day of horseback riding. But make sure the healing retreat is a good match for your particular farm. A day spa client may want to float towards his or her car after a session, still basking in the luxury of relaxation, rather than put the hiking boots back on to return over the hills to the distant parking lot, remembering not to touch the hot wire or accidentally wander into the bull's pen.

Plant-crop based farms will be less of a concern as far as human/animal safety and cleanliness issues, and will provide a peaceful setting and seasonal

A porch overlooking a beautiful farm is a wonderful perk to a day spa.

When a day spa or healing center rents space on your farm

Gregory (Greg) J. Lawless of The Lawless Partnership in Seattle, Washington, contributes points here that you may want to cover with your own attorney, advisors, and with anyone, such as a day spa owner, who wants to rent a building on your farm.

Start with zoning: "A property zoned agricultural and/or residential may not be zoned to allow some types of businesses to operate," Greg says. "Usually the answer can be given by the local planning department." Once you have the correct property location and description, either ask your planning department directly, or just request a general list of what's already allowed in that location. Local planning departments usually can supply you with a list of all the types of operations, from bed and breakfasts to offering massage therapy that can be operated on that property.

Don't give up immediately if you don't see day spa on the list, especially if your day spa is still in the planning stages. Some locations allow homeowners to offer massage from their own home and they can license that business and give it a name. Find out how much can be added to this simple structure to legally create a business similar to the day spa you originally planned. You may find you can operate it after all, but may have to make a few adjustments. Your attorney can help you decide if you'd like to incorporate your business, and if it would be beneficial to allow any other practitioners to purchase a share, making the business legally owned by both you and the others.

Length of lease: While a residential tenant may rent month-to-month from a house just as it is, businesses, such as healing centers or day spas, may be putting a lot into their advertising at the farm's address, and into building improvements, and they need to know that they have a reasonable length of time to be able to operate there to see a return. Also, from the farm landlord's standpoint, a tenant may need changes made in the building in order to operate the business, and if the farmer pays for these changes upfront, then he or she may want the lease amount to include payback, and to be long enough to eventually receive a full refund.

"Most commercial leases are designed to make sure the landlord gets back their tenant improvements," Greg says, "plus a certain rate of return, along with, of course, their rents. For that reason commercial leases tend to be longer. So, the length of the lease really will depend a lot on the number of improvements the landlord will have to make." Of course, if the farm will be gaining larger amounts of customers because of the day spa's presence, this factor may be considered when negotiating any payback for changes the farmer made in his or her building.

Upkeep and maintenance: Who is responsible if the roof leaks or the sound system falters? "Usually the landlord is responsible for the basic, fundamental aspects of a building, such as roof, foundation, walls, and basic code compliance," Greg says. "However, anything associated with the tenant's proposed use is usually passed on to the tenant by either having the tenant pay it directly, or amortized over the course of the lease by higher rents."

Hazard and liability insurance: Farmers certainly want to be protected from losing their generational farms to claims made by their tenants or their clients. The healing center or day spa business needs protection as well, so how do the two of you work together? "Normally the landlord carries hazard insurance on the building," Greg says. "Most leases are triple net, meaning that this cost is passed on to the tenant. In addition, most leases require that the tenant maintain liability insurance, typically for at least a million dollars, and name the landlord as an additional insured.

continued...

CHAPTER SEVENTEEN

When a day spa or healing center rents space on your farm ...cont'd

Usually the tenant will also maintain leasehold insurance on their personal property, and some leases actually require that they do so. People don't realize that not all insurance policies are alike, so you can't generalize on what is or is not covered. So it would be wise to sit down with your lawyer or insurance expert to determine, given the policies you are acquiring, what is and is not covered. If there are gaps in coverage, normally there is a way to expand a policy to cover those gaps."

Fine tuning: "Most agreements need to be fine tuned to the specific situations," Greg says. And the day spa/farm partnership will certainly be unique to each situation. "Often, landlords have a set of rules that have to be followed. In (an on-farm day spa's) situation, you would concern yourself with meshing a farm with a day spa business. So, livestock would be an issue. Odors could be an issue (manure next to a hot tub could take away some of the ambiance). Electrical loads could be an issue. Noise could be an issue. There are a number of examples that should be thought through, depending on what the parties intend, and what are the details of the property."

---

display along with nature's sounds of birds and crickets. "As soon as people arrive on the property and see the fields, they sigh, as it is breathtaking," Ciely says. "People begin to relax immediately. I've had people arrive and start crying, as it supports the much-needed break that most of us yearn for. In the summer, when the lavender is in full bloom, arriving at the farm is like walking into a dream that we usually only see in the movies. Here, it becomes real for a brief time and can be returned to again and again. I believe healing is enhanced when the outside environment supports the inner state one is looking for. The farm provides this opportunity."

Farm animals can, however, also be a huge draw to your customers. Studies have shown that merely watching animals graze can decrease heart rate and blood pressure. Animals' acceptance of us just as we are, and their attitudes and antics, can draw us from our usual inner turmoil, allowing inner pressure to find relief so our bodies can begin to resolve some of their own issues. See the Sidebar, Human Interaction with Animals, for more information on this.

Since local flavor is also an important customer draw for the crowds seeking authentic eco-tourism, a farm spa bonus can include freshly prepared, natural skin treatments, such as facials made with the healing ingredients of fresh farm-made yogurt, fresh-picked, organic pumpkin, or cucumber right off the vine, all of which have been shown to be exceptional ingredients for the skin. Herbal aromas grown on the farm can also add to the overall, one-of-a-kind healing sessions. Ciely suggests that the day spa treatments collaborate with the farm's crops in this effort. Her day spa offers other popular product brands, but lavender is definitely a part of this day spa. "Create (the day spa) as a cooperative effort to enhance both businesses," she says. "The thrust should be the brand of the spa–lavender is our key presentation even though we have unscented & alternative aromas available."

As another example of the farm/health center connection, a farmer in the USA Pacific Northwest rents his barn out to a martial arts school and a group of people who practice yoga, bringing additional rental revenue. The barn offers a big, open and fun atmosphere for these activities. If you're aiming more towards renting space to another business rather that owning the business yourself, see the Sidebar, When a Day Spa or Healing Center Rents Space on your Farm. It offers suggestions on creating a smooth business agreement between your farm and a separate business.

But whether you want to own the business or just rent space for someone else's, stay open to other possibilities if your farm doesn't already have a potential building for the center to operate in. Depending on the services you provide, yurts, portable "mother-in-law" cottages, and even sturdy wall-tents can provide unique atmospheres for healing centers. MaryJane's Farm in Idaho (see Chapter 14) uses wall tents for visiting farm guests, and their tents stay up year-round even in snowy weather. A tent could offer the opportunity to try out the farm-healing center idea temporarily before investing in a permanent situation, or to just expand the farm's offerings seasonally during the most tourist traffic.

While walking amidst Pelindaba's blooming fields, one can almost feel the vibrations of the huge numbers of buzzing honeybees. When operating a day spa in a natural environment, it's helpful to think ahead on many issues, including insect and fly control, which can need different solution approaches than for problems found in urban locations. But keep in mind that day spas operate successfully amidst city pollution, traffic noise, expensive or pigeon-covered parking spaces and questionable neighborhoods. A farm's atmosphere, with a few, or even no, adjustments, may be just what the client's inner doctor ordered and something that matches your farm's goals. Even if you do grow hay, a healing center might be a great option along with, or even instead of, hay rides. As this author was giving a bodycare treatment to a client on her farm, she panicked as the sounds of her farm's haying operation outside penetrated through the open window, drowning out the usual sounds of wild birds and softly clucking poultry. "Hey, John! (amidst excited barking farm dogs) the top @%#*# bails are about to tip, hold on!" The window needed to stay open for a cooling breeze, so I just continued with the treatment. After it was over, the client stretched and smiled. "I grew up on a hay farm," she says. "Those sounds took me back home like nothing else can. That was the best treatment I've ever had!"

Human interaction with animals, including horses. Safety issues, and a new look at the "petting zoo."

In the title *Micro Eco-Farming, Prospering from Backyard to Small Acreage in Partnership with the Earth*, the story is briefly mentioned of a blueberry farmer who allowed his single pet horse to mow the surrounding perimeter of his U-pick blueberry patch. This area wasn't workable farmland, and the mowing kept away encroaching woodlands which would have allowed too many wild animals easy access to the blueberry patch. The horse's manure was mixed with leaves, which eventually produced topsoil so rich with humus around the blueberries, the farmer didn't need to irrigate.

But "Dusty" the horse, brought something more: An experience for the blueberry U-pickers that allowed them closeness to a live horse. He was a memory-making attraction for the farm owners. And although his "job" was unusual, he was seen as part of the farm operation, rather than as an extra addition to a farm just for the zoo appeal. Many experts are pointing out how deprived we are, our children especially, of closeness to nature and animals, and so today, being able to get near farm animals in any way can be a new "crop" your customers find worth paying for.

Many working agritourism farms, including dairies, simply allow the animals their farm raises anyway to be the animal attraction offered to guests, and this is enough. In these cases, visitors enjoy the farm animals whether they see them grazing, being milked, or within a special area specifically meant for human and animal interaction. But, farms that offer a healing element such as day spa services or healing or spiritual retreats may find the added interaction with animals, especially horses, to be an especially attractive bonus to add if they previously didn't keep animals on the farm. As mentioned in Chapter 2, animals can be brought to the farm for special events if you choose not to keep them year 'round. And if you don't farm with horses, have always wanted them, but saw them as a frivolous indulgence, you might be able to find, revealed here, new and at least somewhat profitable reasons to keep them around.

Humans have been partners with horses for thousands of years. Perhaps not until they lost favor as transportation and farm labor did we realize our partnership with them offered something humans can benefit from besides labor. Maybe it partially explains why, in the winter of 2007 with an unsettling war in Iraq and major ice storms hitting much of America, local and national news paused to report that the horse Barbaro, the Kentucky Derby winner, had died. Millions across the nation had sent him cards, flowers, gifts, and goodies including a wreath made of organic baby carrots and a Christmas stocking during his battle to recover from a leg injury that ultimately ended his life. Some even wrote Christmas carols for him.

Horses are also mysteriously popular on Indian Chimney Farm discussed in Chapter 1. Owner Chris Grant raises alpacas and keeps a variety of animals including his own riding horses. He opens his farm to agritourism several times a year, and stated that the very youngest children seem to love the chickens the most. "For just about everyone else," Chris says, "except those who want to start an alpaca business someday, it's the horses. Physiologically they have a huge heart. Somehow it seems as though they show it. They are willing to make a heart-to-heart connection with anyone who is open to this."

Many are familiar with the horse riding facilities specifically designed for helping the physically and mentally challenged. But has our world changed enough to where those who don't fit into the challenged category can still heal in a variety of ways with a closeness to horses and other animals? "The first year we were open," Chris says. "I was told about a boy who was very shy: few friends, few activities, and his parents and mentors were worried. Then he met our horse named Chico. Somehow, this horse melted this boy's fears and he opened up, instantly. I didn't know him, and don't know who he is or where he is. But the experience was marked enough that his coaches and parents thought of it as a turning point."

The dairy goats of GardenHome Farm are used to people. They gather here around their owner, Joan Schleh. Still, customers who come to the farm to purchase farm products are supervised if and when they're allowed to go inside the pens to pet the goats. But even just seeing the goats graze behind their fences makes the purchase of raw goat milk right off the farm an entertaining experience.

Leigh Shambo of Washington State also has had much positive experience connecting non-physically challenged humans, including non-riders, with horses. She founded the Human-Equine Alliance for Learning (H.E.A.L.), which is described as a program that offers services and programs that promote human healing and evolution through the creative and non-violent experience of the horse-human bond. "Our therapy programs, personal growth workshops and facilitator training programs are open to people with any level of horse experience, and no riding is involved," she says. While experienced riders also have found her programs to work wonders for their horse hobby or profession, she has seen great things with non-riders as well. It was her own healing with horses that made her aware of the potential of horses as healers.

"As a young adult," Leigh says, "I earned a degree in Animal Science from Southern Illinois University, and began a career teaching riding and training horses. Then, in 1988 when I was 31 years old, two things happened that set me on a new course. A serious riding accident resulted in a broken pelvis, leaving me with a long period of rehabilitation and significant fear once I did start riding again. And later that year, my mother ended her life by suicide, a shock that prompted me to seek therapy. As I got back to working with and training horses, I found myself undergoing a process of emotional healing that was truly life-changing and profound.

"Soon after, I started learning new horse training techniques: what is now referred to as 'natural training' or 'horse-whispering.' These techniques rely on an emotional bond with the horse, a trusting partnership instead of forceful techniques applied to unwilling horses. Most of this work is done on the ground, without even a halter or rope on the horse. I could not help but notice that the horses were highly attuned to my process of psychological healing and growth.

"On days when I was emotionally 'optimum,' the horses usually partnered right up, without any at-

Goats can be as interested in people as people are in them. Here, a dairy goat named Maude eagerly checks out all visitors who come to the raw goat milk dairy of GardenHome Farm in Washington State.

tempt (from me) to 'make' them do it. On days when I was unwittingly in a negative mood, the horses were reluctant to join up and would resist my efforts in a variety of ways. I could not 'fake it' with horses by telling them I was fine when I wasn't, as people so often do with each other. I also found I actually could listen to the horses, and let them help me become optimum even on my down days. When I let the horses help me get in touch with my feelings, even difficult ones, they became attentive, open, soft and helpful. This is why I say 'Emotional wisdom = horsemanship magic!'

"Working this way with horses was transformative for me, and also for the students I was teaching.

The learning that takes place in relationship building with a horse translates directly to human relationships and life choices. I became increasingly focused on the human psychology involved, and earned my Masters in Social Work. Simultaneously I pursued additional training from other perspectives—the Equine Assisted Growth and Learning Association (www.eagala.org) and Epona Equestrian Services (www.taoofequus.com) in Arizona."

This author observed Leigh teaching her students on a horse farm where she was hired to put on a workshop. She allowed each student to choose a particular horse to work with, and then to enter a ring with the horse who was free within the ring. The student then became quiet, and began to "tune in" with her equine companion.

"Even 'non-horse' people quickly build strong bonds and positive relationships with horses when they are able to practice emotional self-awareness, clarity without judgment, and assertiveness coupled with gentleness and respect. Of course, these very same qualities enhance all human relationships as well! The horses' responses provide immediate guidance to people for accessing these qualities within themselves. There is an old saying, 'The horse doesn't care how much you know, until he knows how much you care.' But, how do we put our caring (for ourselves and others) into practice? In very precise ways, the horses seem to show us the way to translate caring into appropriate action.

"Very quickly, the horses provide proof that who you are on the inside speaks more powerfully than anything else. In a truly amazing and therapeutic way, horses guide people to access higher levels of awareness. Then horses reward the person with

gentle bonding responses. These experiences occur even when we are working with the horse in ways that have nothing to do with traditional riding or horsemanship."

While not all farms who offer casual interaction with horses for the ordinary non-riding farm visitor will choose to offer the depth that Leigh's workshops offer, her knowledge and experience can help explain why a human may have an otherwise unexplainable healing, simply by the approach of a horse. For those with horses who'd like to attract visitors with such workshops, people like Leigh can travel to your farm.

"To give a concrete example," Leigh says about a situation where her workshop became quite valuable to a human participant, 'Sally' (not her real name) was plagued with feelings of regret over her divorce and its effects on her teenage child and her personal

On GardenHome Farm in Washington State, visitor is allowed to touch the goats under the watchful supervision of the farm owner.

Miniature horses are very healing to children. But even at this small size, this horse's young owner carefully monitors him and all children who come near to pet him.

Baby animals such as ducklings are fascinating to non-farmers. But they are delicate and need protection from too much human handling and human-induced stress. Children need to learn that animals aren't toys for manipulation, but have needs that must be attended to. This learning is what brings on the healing aspects of animals for people who are rarely near them.

dreams. Before entering the pen with a horse named Frieda, she expressed the desire to have Frieda give her a sign of forgiveness, an affirmation of her innate worth in spite of her recent experiences. For several minutes, Frieda ignored Sally, and moved away whenever Sally made attempts to approach her. What was Frieda trying to say? This is an example of a situation calling for expertise in both horse psychology and human psychology!

"Very gently I suggested that Sally contemplate what it would feel like to forgive herself. In that moment, Frieda looked toward Sally as if for the first time, and promptly walked over to her, made a small circle around her, and stood by her side! Sally practiced for another half hour. Whenever Sally wanted Frieda to bestow forgiveness, Frieda would move away from her again. Each time she could hold the thought of forgiving herself, Frieda would again join up close by Sally's side. As Sally went back to her world of human relationships, this one experience resulted in greater emotional well-being, increasingly positive relationships with her daughter and ex-spouse, and better success in building new relationships to sustain her through a difficult life stage."

As a farm offering agritourism that leans towards healing, your choices are many as far as how your guests can experience horses or other animals in new and positive ways. If appropriate for your situation, consider adding the occasional, deeper healing workshop by inviting a practitioner such as Leigh of H.E.A.L. to offer workshops, giving your community and tourists something brand new to experience.

Others who work with horses prefer to offer more traditional animal experiences, such as pony rides and the experience of a horse-drawn vehicle, or even weaving in living history lessons at the same time such as what the author has done when school children attending a living history re-enactment gathered on her farm to see how a pony is hitched to a cart. And, some enjoy offering traditional horse experiences, but with miniature horses instead, offering the experience of a real horse without the higher danger and intimidation that can be caused by the huge size of standard horses. Jim Walsh and Kathi Dunphy, owners of Minihorse Farm B&B and Cottage Rental in St. Martins, New Brunswick, Canada, enjoy tending their B&B, cottages, and large organic gardens from which they produce farm-made jams

CHAPTER SEVENTEEN

and jellies for sale. They also raise registered American Miniature horses on their eight acres, and offer farm guests a chance to experience cart driving and for the very smallest, a saddle ride.

Safety and liability issues

If you own a wool ranch, dairy, or other farm that incorporates animals as part of its crop production and can't viably tame them, simply allowing guests to see the animals from a distance can be a very enjoyable experience as they observe the animals in their natural farm environment. A hybrid of this could be a dairy or sheep ranch, for example, that keeps guests at a distance from their regular, less-tamed farm animals, with just two or three especially tamed and clean animals nearby and supervised when guests arrive. Children can be so removed from nature in this century, that it can be more valuable, enjoyable, and safe, for a trained human companion to be with the animals available for guests to get close to. Kids can be shocked that there's a give and take with animals, that the creatures aren't objects of entertainment and control as are computers and ipods. Seeing another human interact properly, and give directions on how to get an animal to trust you along with safety measures, can ultimately offer children a safer and far more fulfilling experience than their usual attempts at impersonal instant gratification.

Liability insurance for horses, unfortunately, can be too prohibitive for some to add them to an agritourism package. One owner of a cattle ranch who turned to agritourism found that just about every endeavor, even allowing guests near their cattle, was quite simple to insure, but had they had horses, insurance and on-farm expenses for safety measures would have been too costly. Even if the horses are small, and won't actually be ridden, they have kicking habits and teeth that can bite in a more harmful way than ducks and sheep. When a child who doesn't know horse behavior enters a pen with a pony and kneels down to tie her shoe, her head is dangerously close to kicking hooves. I've seen friendly miniature horses enjoy the attention from the children so much, that the herd leader thinks she should have it all to herself, proceeding to "gently" kick the other horses away. Those kicks, even if playful, will have a different impact on their own kind than it will to a child's head.

But Chris Grant, the owner of Indian Chimney Farm who allows visitors to see his horses on a more casual level during other agritourism events, feels he solved the kicking and biting problem of his horses. He keeps them in a pen separate from people, and only the horses who are interested come to the edge and allow visitors to touch them from the other side, which is also less stressful for the animals than being put into a petting area to be reached for whether the animal is "in the mood" or not. Chris says that only certain horses of his will come greet the public. "All horses kick and bite," he says, "so we keep them in

Leigh Shambo attracts visitors to a horse farm by giving intriguing workshops on how to develop a deeper relationship with horses.

A student at Leigh Shambo's workshop uses Leigh's techniques to develop a strong human-horse bond.

micro organisms mostly beneficial, including intestinal microbes such as acidophilus? Studies suggest when this is the case, the pathogens that cause disease can't take hold as easily in either the animal nor the human when beneficial microbes are in abundance.

Since you can't control the health of the children visiting or ensure that their immune systems are healthy enough to withstand any disease-causing pathogens, however, it is suggested that you take maximum safety precautions. Below are a few suggestions to follow as a starting point. Be sure to follow your own local health regulations and updated suggestions from your own region, as even different localities have higher and lower incidences of pathogens being spread by farm animals, and new information and legal restrictions continually become available on how to handle cleanliness to maintain health. Your local Department of Health and local agriculture extension agent may be of assistance, and if so, avail their help and advice. They're probably very knowledgeable in this and have some practical suggestions that may help to both to keep everyone safe and healthy, and to show that you followed the best advice available in the event of a lawsuit. If you're hosting school children, discuss cleanliness with the teachers ahead of time and make sure they agree to supervise proper hygiene.

Hand-washing

Offer a hand-washing area(s), both in animal petting areas as well as outhouse or bathroom areas, with liquid soap and paper towels. You may want to check to see if composting the paper towels is viable.

Supervise children in washing properly. They can find this to be very fun! Have them wet their hands thoroughly under running water, then lather with soap. Once lathered, briskly rub hands together

the round pen. That fixes the kick part. As for their mouths, the good ones will only bite if they think your finger is a carrot. Bad idea. Their jaws are very strong, so a little 'nibble' turns into a big 'ouch'. That hasn't happened here, but we don't allow the public to feed the animals. That's what makes them nippy."

Bio-safety, germs & disease

There is much change and debate going on, also, about bio-safety, germs and disease. Every country and county has their own viewpoint and safety suggestions and regulations. A study in Europe found that children raised on farms and indirectly digesting soil and other "dirt" developed superior immune systems. Others point to incidences where pathogens that passed from soil or animal to human caused death. It easily could be argued that it all depends on the balance of micro organisms within the humans, soil and the animals involved. Are those

CHAPTER SEVENTEEN

for a minimum of 20 seconds (for younger kids, time a fun nursery rhyme that's at least 20 seconds, perhaps one that pertains to farming, and have them rub their hands for the amount of time it takes to sing that rhyme.) Remind them and demonstrate to them that they should cleanse palms, backs of hands, fingers from bottom to tips and nails. Rinse well under running water. Dry with paper towels (not on clothes) and if water taps are involved, turn them off with a paper towel. Toss the towels into the garbage or a compost collector.

Offer sanitizing gel or wipes. Again, there is debate on whether over sanitization is beneficial or actually harmful. Depending on where you are with this debate, choose either the strongest commercial ones available or a natural alternative proven to be as, if not more, effective. Some more natural alternatives may offer the benefit of killing pathogens while being less destructive to beneficial bacteria.

Make sure all your animals are very healthy, including their intestinal flora, and that the area where they are held when people may touch them is fresh pasture or otherwise a clean area. Portable pens moved around a pasture can be useful for this.

If very concerned, suggest kids wear boots while at the farm, then change back into regular shoes, bagging the boots for cleaning later by an adult, and washing hands after the shoe change.

First aid & CPR

If your local ag-extension, local authorities, attorney, or even you, yourself, come to the conclusion that a first aid area and a person nearby who knows CPR would be required or advisable for the numbers of people you're expecting, how will your guests find them and who will be in charge? Some festivals set up a special area where someone is always waiting nearby with first aid and CPR training, or who is at least standing by to contact such a person via cell phone who's occasionally in another area of the festival. Signs in various areas of the farm, including the first entrance, point the way to the First Aid Area. ❧

Seasonal Agritourism Events 18

"**D**o not underestimate the entertaining value of your farm. People are dying to experience the farm, and they need to see it! Folks need to get in touch with their food," says Brad Stufflebeam, owner with his wife Jenny, of Home Sweet Farm in Brenham, Texas, USA. They operate a Community Supported Agriculture (CSA) farm, a farm consulting and design service, and yearly farm events including an irresistible heirloom melon festival. And while the Stufflebeams engage their farm customers in very unique, on-farm entertainment activities, it's good to remember that the traditional roadside stand and the newer CSA on-farm pick-up programs are important forms of ongoing agritourism on their own.

"We offer agritourism every week as our CSA members pick-up at the farm," Brad says. "We also host quarterly events that are centered around the farm's seasonal availability: Spring Planting Festival, Tomato Fest, Melon Fest and our Fall Festival kicking off the pumpkin season."

Home Sweet Farm's agritourism activities progress and change with what works and what doesn't, as all wise farms operate. Besides the farm's seasonal events, agritourism has included the regular CSA membership produce pick-up on Wednesdays, and an innovative Sunday program the farm created where farm members who paid a yearly membership fee harvested their own crops between 11 a.m. and 5 p.m. The farmers designed the program so that non-members may also join for a per-visit fee.

"We originally wanted to incorporate a U-pick so that others could experience the farm. Of course, it helps to have the harvesting spaced out during the week," Brad says about the Wednesday and Sunday harvesting schedule. "We have found people are interested in getting in touch with their food. We give our CSA members the first fruits every week, but when a bumper crop is coming on, we like to share that with others." Home Sweet Farm produces some tantalizing fruits, such as heirloom melons like "Tigger," the highly perfumed melon with the electric yellow rind splashed with bright zigzag red stripes. "For us," Brad says about customers coming direct to the farm, "it beats going to the farmers' markets. People come as far as Houston and Austin just to experience the farm."

Allowing success to breed more success while growing gradually

"Our farm is 22 acres of diverse soil," Brad says. "We started our first CSA season with a one-acre garden planted intensely with a variety of produce, focusing on heirloom tomatoes, peppers and melons." Word spread, and so did crop varieties and farming events.

While some farmers evolve into agritourism as an afterthought to their farms, Brad and Jenny thought differently. "We always intended to have people come to the farm, because of our interest in education and community development. However, friends and associates encouraged us to do it sooner than we planned. They helped us realize that seeing a real, small, family farm starting up from scratch was a major educational opportunity, not just for our farm customers, but also to encourage and help grow new farmers. We had an abstract business plan, with broad details. We knew we would incorporate festivals, seasonal pick-your-own and a farm CSA pick-up (that's agritourism every week, and a majority of our farm sales)."

Brad and Jenny were not new to innovative and sustainable farming when they began. They had previously owned and operated the first 100-percent organic nursery in Collin County, Texas, USA for almost 10 years. Brad was operations director for World Hunger Relief, which operates out of Texas, addressing the needs of domestic and worldwide hunger. He ran a Grade A raw goat dairy, raised hens, and produced award-winning honey. He has trained others in sustainable agriculture development for third world countries through World Hunger Relief, Inc. This non-profit was chartered in 1976 by real estate developers Bob and Jan Salley. Three years later, Carl Ryther and his family returned to Texas after 17 years of agricultural missions in Bangladesh (formerly East Pakistan). The Salley's invited Ryther to join World Hunger Relief and together they developed a program to train individuals to address hunger around the world. Their interns now work for various international organizations around the world promoting sustainable food production and sustainable local economy.

Those past achievements and experiences in production, marketing, and even farm events have obviously served Brad and Jenny as they continue to develop their family farm. "We have had Farm Days in the past while we managed the World Hunger Relief Farm in Waco, Texas, along with other festivals related to our nursery/plant farm business that we owned for nearly a decade in North Texas," Brad says. "Our special events are always key elements to

Home Sweet Farm
www.homesweetfarm.com

getting people out to see our work and promoting sales."

Education is also a part of their festivals. "We have tried to make the events educational by inviting guest speakers, giving tours, serving food and having live music, which allows people to meander and just relax. Our upcoming events are designed to get families more involved by celebrating the season. This spring, once the major planting is done, we have invited members and others who are interested to come out for a workday in the garden."

Brad and Jenny also devised a plan to allow guests and members to feel the joy of planting crops. "We are going to let the families plant one bed of tomatoes," he says, "so they can gain an appreciation for the work, and then invite everyone back out to the farm in July to celebrate at our Tomato Festival. The festival will be a day of fun and farm tours along with a sampling and survey of over 30 varieties of heirloom tomatoes. That's worth celebrating, and everyone seems to be excited about the program."

Being a family- and visitor-oriented farm, Brad and Jenny are prepared with favorite autumn crops for guests as well. "Our Fall Festival will be centered on our U-pick pumpkin patch," Brad says. "We grow unusual heirloom pumpkins, winter squash, sunflowers and gourds to make things different. We really focus on a lot of diversity. People want different. And we want to turn our community and children on to local food."

Photography by Jennifer Stufflebeam

Photography by Jennifer Stufflebeam

Photography by Brad Stufflebeam

(Above and Left) "These are from a 'gleaning' that we organized with a youth group from a local church at the end of our fall season. We harvested 18 cases of greens (mustard, collards, salad mix) along with radishes and turnips donated to 'Faith Mission,' an organization that provides food and housing to the homeless in our county. It was greatly appreciated."

– Farmer Brad, HOMEsweetFARM,
www.homesweetfarm.com
Photography by Pam O'Brian.

CHAPTER EIGHTEEN

> *Do not underestimate the entertaining value of your farm. People are dying to experience the farm, and they need to see it! Folks need to get in touch with their food.*
> – Brad Stufflebeam, Home Sweet Farm

Diversified promotion

The events no doubt boost farm promotion, and when it comes to spreading the word about Home Sweet Farm, Brad is certain of the course that works best for him. "Nothing works better than word-of-mouth," he says. "People talk about quality food, and unusual heirloom vegetables." This word-of-mouth excitement really helps spread the word about Home Sweet Farm. Also, as with many sustainable farms that allow visitors, a sense of ownership and involvement draw out loyalty from customers. "Our members see the farm as their own," Brad says, "and are proud to share their experience with others."

Brad and Jenny also market online. "The Internet has been a big tool for introducing ourselves to potential members," Brad says. "Listing our farm with localharvest.org, allorganiclinks.com, foodroutes.org, and csacenter.org has helped new people to find us. Having a simple, well-designed website that we keep up to date has been very important." To lead people to the website at www.homesweetfarm.com, Brad and Jenny have found some valuable methods. "We focus on well-designed advertising presented on our website and e-mail list, along with (listings on other websites mentioned above) and our monthly e-newsletter. One special way of reaching out is through area Yahoo Groups. We have found online communities in our region in which to connect to others who are interested in quality food."

Traditional print media has been helpful to Brad and Jenny as well. "We also leave cards and brochures at local businesses where we think our mem-

bers might shop, like the local health food store, and a teacher's supply store," Brad says. "Teachers are often looking for wholesome, unusual and educational field trip possibilities."

How about networking with other organizations for farm promotion? "We do it all independently," Brad says, "and market aggressively on the Internet, and by word-of-mouth (all for free, basically). As we develop more infrastructure, we look forward to working with our local Chamber of Commerce in the future, which does an excellent job directing tourists around our historical county (#2 in the state for tourism). We also plan to do educational workshops with TOFGA (Texas Organic Farmers & Gardeners Association) next year to help promote local and sustainable agriculture."

The Stufflebeams allow customer feedback to help guide them in what produce is the most popular. "During our Tomato and Melon Festivals," Brad says, "we sample many varieties of heirloom vegetables, and I always take notes on the favorites."

Pricing and operating on-farm events

While pricing methods evolve and changes from year to year, Brad shares his recent pricing scheme. "Our Festivals are free, more for our members and to promote the farm to the public. We have produce and plant sales in the spring, offering heirloom vegetable transplants. Our pumpkin patch is $5 per pumpkin, and includes a hay ride to the field. We also incorporate children's activities during our festivals like our 'ladybug release party' every spring—we charge $2 to help cover the costs."

As the farm progresses, its owners tweak and add new agritourism opportunities. "Technical Farm Tours are also available for $50 which includes lunch," Brad says. These were created for others interested in developing their own small, family farm. "In the afternoon," he says, "we open the farm to the public, and invite area farmers, co-ops, and those involved in farmers' markets and restaurants to set-up informational booths to inform others

"This is a scene of our girls and others releasing lady bugs over our spring fields. "The Lady Bug Release Party" is a part of our annual spring farm day, 'Spring Fest,' which is always a big hit! We order 100,000 lady bugs for the kids to set free. We also have extras for them to take home."
– Brad Shufflebeam, Home Sweet Farm
Photography by Jennifer Stufflebeam

about our regional food supply. The afternoon also includes a Family Farm Tour, guest speaker, petting zoo and other children's events."

Brad's advice for hosting events includes making sure there is enough staff on hand, and being prepared for lots sales of on-farm products. "The biggest mistake is being short staffed during events," Brad says. "You can easily get members involved to volunteer and help out. It's important to get our community of CSA members involved. After all, they consider this to be their farm, and they enjoy helping and showing folks around." Brad feels that even more farm produce could be sold at the events than what he currently offers. "I wish we had a lot more planted to offer for the festivals. We are passing up a lot of sales, and are having a hard time keeping up with the demand for small farm products."

To help with food safety issues, Brad and Jenny have found a way to network with others for mutual benefit. "We always include a reputable local restaurant featuring our farm-fresh food. They prepare everything in their kitchen, and handle food properly on the farm."

Like many successful eco-farms today, Home Sweet Farm is in continual motion from year to year, with a successful balance of solid programs that create a secure financial foundation as they experiment, innovate, weed out what doesn't work, and expand on what does.

The agritourism payoff

The sweet melons and other Home Sweet Farm produce have gained great popularity and a loyal customer following through the farm's agritourism. "Farm events provide immediate on-sight sales, but the PR is invaluable! It makes a great story, and newspapers easily pick-up the news. In fact, they are dying to report good news… What's better than a small family farm in the local community? The value down the road is immeasurable. We go out of our way to include other farmers and regional non-profit organizations working to promote sustainable/local/organic agriculture. Our farm cannot provide everything, and we want to network with others to help make local food more convenient for the customer. We definitely learn a lot from having folks out to the farm, and our girls (their two daughters) consider the farm events a big party. We really enjoy hosting special events, and sharing the realities of our farm. The public appreciates it and needs to experience it."

When asked about the idea of superficial agritourism entertainment versus "real farming," Brad had a quick answer. "We are not looking to attract tourists who do not understand the farm or environment. We promote the fact that we are an authentic, small family farm. That is the draw. You cannot fake it, and we do not apologize for it. We keep cardoon growing for the kids, as a trap plant for aphids and other pests, which becomes a ladybug breeding ground. We can show kids the entire lifecycle of the ladybug there, from nymph, to mutation to ladybug.

CHAPTER EIGHTEEN

It's incredible watching their excitement as they see a balance in nature. We love to see people coming together. It is amazing how food connects people, and our farm is becoming a center of community larger than we ever anticipated. When our visitors appreciate us, it makes our work that much more rewarding. It really encourages us personally and keeps us going. That's what it's all about: our committed farm friends who celebrate our farm being here. How can you fake that? This is for real, and people need the reality. It provides an emotional connectedness for them, not just simply entertainment."

Heirloom melons as an agritourism draw

Here's an example of choosing a focus crop besides the ever-popular pumpkins and corn mazes as the product for a special harvest agritourism event. Here, we'll explore the idea of a summer U-pick vintage melon patch as an agritourism draw, with inspiration from the melon patch of Home Sweet Farm

"There is not much opportunity for folks to realize how they're a part of agriculture," Brad says. "What better way than for a family to explore a melon patch? The kids really light up, and food becomes fun."

Much has been said about the agritourism opportunities for pumpkin patches and corn mazes. Throughout this book, we've seen festivals and gatherings revolving around lavender, flowers, apple harvests, theme gardens, artisan cheese, and many other crops that became wonderful focuses. But for those with diversified fruit and vegetable farms, it seems hard to go wrong with a sweet summer melon patch. And perhaps an heirloom one at that.

"Heirloom melons look different, that is for sure," says Brad. "The aromas and flavor cannot be matched with conventional commodity produce. Many melons could be a dessert."

Indeed, melons are the candy of garden fruits and vegetables. A vintage agritourism melon patch can be a small farm's candyland. There are melons with flavor hints of mango, pineapple or peach. There are vintage favorites grown by Thomas Jefferson and French melons with ambrosia-like scents. The Kiwano melon has flesh that resembles lime green Jell-O; and then there's Tigger, that taste-test winner first discovered in Armenia that Brad and Jenny now grow on their farm. Its flesh is so aromatic, that cutting open one Tigger perfumes an entire room.

The history and flavor of melons can be an intriguing draw to your customers. It seems melons have been with us since the beginning of time. Thousands of years ago, African Bushmen relied on wild watermelons to provide living canteens of uncontaminated water. Wild relatives of our current watermelons and other familiar melons still grow in Africa. But the flavor, variety and geographical range of cultivated melons have certainly come a long way. It is believed the seeds of the sweetest were saved over time and appear to have been cultivated in Egypt at least 4,000 years ago. Around 3,000 years ago, they reached beyond Egypt via trade routes, and most likely arrived in Europe during the Greek and Roman times when North African Moors brought melons to Spain during their occupation from 711 to 1492. Melons and cucumbers were planted in Haiti when Columbus landed, and early European colonists and African slaves brought more melons, including watermelons, to North America. Native Americans took them eagerly into their gardens. But it was not just the flesh that was valued historically. Watermelon seeds also gave nourish-

A neighboring dairy farmer, Darlene Barrett, provides a petting zoo for a festival held on Home Sweet Farm.
Photography by Pam O'Brien

ment, especially to Africans and Chinese, over many hundreds of years. In more modern times, Chinese medicine and recommendations from Edgar Cayce, who widely promoted holistic healing, recommend white watermelon seeds and a tea made from them as a remedy for mild ailments. Even the rind is sometimes pickled for food or used for healing.

By the 1800s, a rainbow of melon varieties graced gardens and filled farmers' patches. But like many food crops, melons would eventually undergo hybridization and commercialization in the 20th century. They were bred for centralized agribusiness, long-distance shipability and for growing and harvesting conditions that suited larger, commercial operations. When this happened, many of the vintage melons of yesteryear went by the wayside. Conventional hybrid watermelons as well as the netted melon Americans have come to call cantaloupe, and the honeydew, became the norm in supermarkets, with a few novelties showing up here and there. But there are hundreds of worthy melon varieties available, and more being rediscovered or re-bred. Locally-grown heirloom melons make an especially inviting gourmet market crop, as they are rare in supermarkets. They need to be vine-ripened for maximum sugar content, which often takes careful hand-harvesting at various spells over the growing season. They do not sweeten much, if at all, after being picked. However, some can be stored for weeks or even months, and during this time, juiciness, texture and aroma may continue to improve.

Heirloom melons have also impressed Josh Kirschenbaum of Eugene, Oregon, who is the product developer of seeds and green goods for seed companies, including Abundant Life Seeds, which tests and distributes heirloom varieties. "This past summer, we trialed over 75 different melons and over 80 percent of them were open-pollinated," he says. "There were quite a few impressive varieties with probably just as many reasons why they were impressive. One thing that stood out was that the varieties touted as being early-maturing varieties did

indeed ripen at the same time as their hybrid counterparts."

Those who trial seeds for specialty markets know that the varieties must be top quality to retain loyal customers. Vintage melons, with their great variety, are attracting customers quite well. "Another trait I found impressive about several of the older, open-pollinated melons," Josh continues, "was the unique shapes, sizes, and colors. I think that for some folks, all they see are the standard melons found at the grocery store. For those that are looking for something unique, there really are great choices out there."

But what about practicality? Are they as "tough" as hybrids? Apparently many of them are. "Something else that I found impressive," Josh says, "was that several of the older, heirloom types were quite disease resistant."

An all-time favorite summer treat, certain melons can also be pickled and preserved. At Cedar Hill Farm in Rice, Minnesota, fresh honeydew slices are simmered in vinegar, allspice, orange peel, nutmeg, cinnamon, cloves, sugar, and pectin to make a pickled melon, which they sell over the Internet. And the varieties that store naturally well into the winter extend the delights of these fruits.

Sharing melons and selling them to others

When this author grew her first specialty melon more than 15 years ago on Island Meadow Farm in the Pacific Northwest (where local vine-ripened watermelons are rare due to cool summers), I sliced it in half, and tested one of the halves on a very conservative friend. It was a small, yellow-fleshed watermelon. "I kept it in the refrigerator for a week before I could touch it," she says. "It seemed so strange to see a yellow watermelon. Once I ate it, it was so delicious."

Since then, most folks have opened up and come to expect wonderful things from the world of heirloom melons. "We have been growing melons now for only six seasons or so," Brad says. "We started, naturally, with an interest in old varieties; our first was the now popular Moon and Stars watermelon."

Moon and Stars is a large, deliciously-flavored heirloom watermelon with dark-green rind speckled with bright-yellow dots that resemble stars in the night sky, and often at least one larger yellow spot (a moon). It has come to be better known by the public since it first reappeared in rare seed catalogs near the end of the 20th century. Since its rediscovery and introduction to the general public, citizens of our country seem to want even more melon surprises. "Most everything we grow is unusual, and educating the public is the center core to everything we do," says Brad.

Jere Gettle travels the world seeking fine heirloom seeds. Jere is the discoverer of the Tigger melon and now sells many vintage melon seeds to growers across the globe (see Resources below). He says that the key to drawing customers to newer crops is variety. "Customers buy more and try more when they see the wide variety available," he says. Jere also suggests offering recipes, serving ideas, and free samples if you can. "Once people try many rare varieties, they want more!"

Choosing melons for your patch

When selling to customers, this author and others find that a combination of supplying familiar favorites from seasons past, along with new surprises each year can help bring customers back again and again, while doubling as testing ground for new varieties that may eventually become the familiar future crop staples. The initial choice of possible melons for your summer patch is huge. A book by Amy Goldman (see below) and open-pollinated seed catalogues can give you a good start in choosing. Here, other growers are happy to share their favorites to help you make your decision. Brad and Jenny, in fact, use the services of an expert to help make their choices. "We try new melon varieties every year," Brad says. "This year's additions where selected with the help of our seven-year-old daughter, Casaba.

Farmers' daughter checks out melon patch on Home Sweet Farm.
Photo courtesy Home Sweet Farm.

They include Golden Beauty, Charentais, Banana, Kansas, Orangeglo, Tigger (how can you pass that one up), Ogen, Golden Midget, Boule D'Or, and Black Diamond Yellow Flesh (we expect this to be a hit… Black Diamond Red is famous around here)."

The Charentais that Brad mentioned above is also a favorite of Jere's and many others. It is considered by Amy Goldman, author of *Melons for the Passionate Grower,* to be the melon favored above all others by the French, which she describes as having a "divine scent and ambrosial flavor." The Charentais originated in western France's Poitou-Charentes region around the early 1900s. Gourmet and heritage seed collectors have found it grows well anywhere in the United States that other melons grow. Jere describes it as "top of the line when it comes to marketing, and the two- to three-pound fruit sells for about three dollars a pound at many markets (double to four times the price of regular melons at the time)."

Jere is also very impressed with Boule d'Or (Golden Perfection). "This is my favorite honeydew type melon that was once a market melon in France and now is a winner for classy markets. The flavor will win you over to heirloom varieties with the first bite!"

Piel de Sapo (Toad Skin) is a very rare, white-fleshed melon Jere likes that is from Spain with exceptional keeping quality. About eight inches long and oval, it is a late-season variety. "This variety can be sold over a long period," says Jere, "and is good for autumn sales. With its fragrant, mellow flavor that is richly sweet; it can keep until Christmas."

When choosing a single crop for a special event, melons versus pumpkins, for example, consider the situation of the families who will be your customers. Pumpkins are harvested when most people in your own community are at home, returning to the school year routine and plugging back into their own community after a summer with time spent away. Schools themselves are great places to market your pumpkin patch. Melons, on the other hand, are a summer crop. Members of your own community may more likely be scattered with preparations for hosting family reunions and taking family vacations outside the area. Cater to your possible customers' situation with your promotion. A melon patch can be great fun for those who are on vacation in your own area, as well as an event your local community members might want to bring their visiting relatives to see. A better place to promote a melon patch beyond your own website and farm CSA might be posters put up at local tourist attractions to draw the visiting crowds, leaving schools for the pumpkin patches. If you already have a pumpkin patch, also, it won't hurt to hand out "Mark your Calendar" fliers about your summertime crop festival, and offer a place for visitors to sign up for e-mail or postal mail reminders of the event.

For resources for rare and specialty crops, including melons (see Resources). ❧

Part V: Epilogue

Epilogue: Agritourism around the world

Agritourism around the world: What we can learn from each other. Ideas and challenges from the USA Pacific Northwest, the French Countryside, the Island of Crete and the tip of South Africa.

As mentioned in Chapter 1, within the next decade, seven billion tourists are expected to explore deeper into their own regions as well as far away lands. What about the places that will host them? Already, some of the most popular tourist destinations are straining to maintain their true identity while resorts and casinos try to push their way in. But there are answers, and eco-tourism, including the rise in agritourism, is one answer that can add revenue to local economies in a manner that helps them maintain or even upgrade their local voice and flavor. It spreads visitors out across the globe, drawing them away from congestion to places not on the regular map of destinations.

Even if this prediction of exceedingly huge numbers of roaming citizens proves less than previously thought, travel is a long-established preoccupation of humankind. It started with trade routes and explorations among neighboring tribes and kingdoms seeking adventure, new goods and knowledge. We've already described how historically, travelers have long depended on local farms as places to stay on their way to visit new areas, or even as they crossed the country to seek their fortune. Even in the wake of USA's 9/11 tragedy, travel soon resumed and bypassed record numbers, and when fuel costs make long-distance travel more difficult, tourism to closer destinations picks up. Travel will no doubt always be part of the human condition.

Agritourism is spreading worldwide, and we can all learn from each other as we strive to maintain our regional uniqueness and locally rooted economy, and to benefit financially from agritourism while siphoning revenue from the affluent to those who, traditionally, were on the lower end of the pay scale. According to World Tourism Forum for Peace and Sustainable Development (WTFPD), "In a variety of ways, tourism has the potential to contribute decisively to the development of a world that's more peaceful, while diminishing the rift between haves and have-nots, and promoting awareness of the need to rely on sustainable productive processes." Heifer International (HI), a non-profit which works to help those in poverty help themselves through sustainable agriculture projects such as raising honeybees, dairy goats and dairy cows, takes visitors on study tours to remote and unique ecosystems. These tours reveal locals working to improve their lives while preserving the planet. Such study tours, HI says, will take participants beyond their ideas about poverty and hunger, and show the power of transformation and the human spirit.

But let's also heed a warning. Among those ancient tribes and kingdoms that sought adventure, new goods and knowledge, it's no surprise to anyone that there were those seeking goods to steal and

Agritourism is spreading worldwide, and we can all learn from each other as we strive to maintain our regional uniqueness and locally rooted economy, and to benefit financially from agritourism while siphoning revenue from the affluent to those who, traditionally, were on the lower end of the pay scale.

victims to add to their empires. Not a single race or continent can claim to have none of this in its history. As agritourism continues to grow in popularity, it sometimes finds itself at a fork in the road like all things that reach a certain distance. Taking one fork has a better chance of contributing to the peaceful and sustainable world described by WTF-PD. The other could allow those who aren't qualified, yet have current power in prominent positions worldwide, to twist agritourism on par with some multinational corporations. I'm speaking of the ones who propose to save the world from starvation via the serfdom of farmers and extinction of nature as we know it, in replacement for genetically modified patented crops and those under obedient contract to

grow them. Governments can and have been proactive and of great benefit to farms and agritourism in some instances, and multinational groups, such as the Slow Food Movement, have been powerful instruments in saving local tradition and regional economies.

The fork we want to avoid, though, is the takeover of agritourism by non-farming, non-visionary government control, or the creation of big business and multinational, corporate-modified farms, (maybe even patented!) operated by non-farmers who know nothing of nature's holistic complexity or connections to actions we take now and their result down the road. At the other end of the polarity is usually something that's feeding such empirical entities, which is the complacency of large enough numbers of citizens, rather than the power of the human spirit described by Heifer International.

For this epilogue, we'll focus on four locations which are very unique from one another socially and geographically. We'll go from the USA Pacific Northwest to the French countryside to get a glimpse of the potential of government support for farms and agritourism, and then to the island of Crete and the tip of South Africa to look deeper into that power of the human spirit as it applies directly

Tourists driving along country roads during an agritourism event enjoy seeing cattle graze the naturally fertile pastures of Skagit Valley.

A view seen by visitors to the Skagit Valley tulip fields.

to agritourism as these locations face both the abandonment of support by larger entities, and the threat of big entity takeover.

The Fertile Skagit Valley, USA Pacific Northwest

Let me share some revealing news from the Skagitonians to Preserve Farmland (SPF), a group dedicated to preserving and enhancing Skagit Valley's farming industry, one of the most fertile farming valleys in the USA. Here's a quote from a past SPF study:

"Tourism generates substantial revenue for Skagit County annually. Every April, for example, between 300,000 and 500,000 people attend the annual Skagit Valley Tulip Festival, an event made possible by Skagit County tulip and other farmers. It has been estimated that this event generates an annual direct economic impact of $15.3 million plus secondary impacts of $6 million and $0.9 million in labor income. This activity and revenue would not be possible without farms and farmers.

"Agriculture is a central necessity for the success of the local tourism industry, creating an authentic, identifiable sense of place that can be appreciated by tourists year round. Fewer and fewer communities can boast this local, rural authenticity. Agriculture puts a face on the community.

"Owners of bed and breakfast (B&B) operations commented that even a small, 3-room B&B operation, for example, may easily serve 800 to 1,000 individuals each year. A slightly larger one would easily serve 2,500 people. An overnight stay will almost certainly result in the purchase of dinner at a local restaurant, as well as other spending in the community. These revenues would not be possible were it not for the scenic values provided by our farming operations.

The group also noted that tourism is more of a burden than a benefit for most farmers. "Traffic jams occurring during tulip season, for example, can be a substantial inconvenience. Yet the economic value of this tourism is a definite benefit to the community and results, at least in part, from the presence of an authentic agriculture industry."

The Skagit Valley which SPF reported on above is filled with fields of tulips, daffodils, irises, grain, vineyards, apple orchards, nut groves, hay fields, cole crops, small diversified farms, and many others

The French countryside, historic castles, and farmland are preserved and mingle beautifully in France, providing an attractive combination that includes agritourism.

As agritourism grows, a portion the revenue it draws in needs to reach the farmers. That seems obvious enough. But it doesn't always happen. It can reach them directly by the vision and innovation of the farmers themselves, of course—that's what this entire book is about. But governments and advocacy groups also can contribute to their farmers in even more ways: perhaps by creating a pool that can be dipped into during natural emergencies; brandings of regional flavors; offering shared advertising of local products in a manner that raises pay to the farmers (a simple, higher price for the product doesn't always reach the farmers); a system that helps with liability concerns for on-farm visitors; training farmers how to operate their farm B&Bs or better handle roadside stands and take in revenue from those traffic jams during harvests that attract so many tourists.

The SPF is already working towards a number of ideas like these, and others may like to glean their ideas. Their website, www.skagitonians.org, offers a wealth of practical solutions they've been successful with so far. These include alliances with local hospitals to sell local farm products direct, a land trust that raises funds to pay farmers for development rights so the land remains farmland into the future, and public educational programs to maintain support for farms by non-farming citizens, such as their regional radio station, "InFarmation Radio," that broadcasts regional farming news.

The French countryside

As a rancher's daughter who grew up reading children's books of talking horses and watching Foghorn Leghorn on TV (an old American animation about a ridiculous talking rooster), I learned quickly that food and farming were far different than many Hollywood or storybook renditions.

including grass-fed dairies where passersby stop to marvel at the cows grazing in the sunshine. Historic red barns and white chapels were once more abundant, though a number have fallen to lack of funds to keep them maintained, while track housing can be seen approaching in the distance. While watching the news one winter, I heard a farmer tell about how migrating snow geese had wiped out his green pastures. This dairy farmer was one who went the distance to farm in a natural way, with open pasture, and open space that drew many people to visit the beautiful countryside. But when asked how much the replanting of his fields would cost, and the emergency hay he had to bring in, the cost was astronomical. Non-organic feedlots may have been a better economical choice. Was there any help from the county for this natural emergency? Maybe something from that $15.3 million? No, he was on his own to foot the bill. Let's hope that type of unnecessary hardship is changing. The SPF sponsored the above quoted study to show the value of their local farmers, and have since moved forward with that information to make things better for their region's farmers in a number of ways.

That is, until I traveled as an adult to France. Older French women with scarves tied beneath their chins really did open their stone-house wooden window shutters adorned with flowered window boxes and say, "Bon appetit!" to French bakers wearing berets and pushing wooden carts of baguettes down winding, cobbled streets. Endless, rolling, green hills really were dotted with content dairy cows grazing in the sunshine or chewing their cuds towards evening. The geographical namesake of the Toulouse

It's hard for tourists to resist the pride in historic preservation, as in this restoration near Paris of the peasant farms developed during Queen Marie Antoinette's reign in France.

goose, Ruen duck, and even the nearby islands of Jersey and Guernsey make travel here for those seeking deep roots in food and farming a nostalgic movie or storybook dream come true.

Agriculture dramatically shapes the environment. In much of France, the agriculture has created a stunning environment. France is rich in authentic tradition and they know how to make their traditions pay in the form of tourism. Yes, of course there are tourist traps, but there is deep, authentic tradition in France that draws people in huge numbers. "France has a tradition of métier, pride in a lifelong career, be it plumber, chef or diplomat," says Chef Sally McArthur, who travels between France and the above mentioned Skagit Valley farmland to put on French culinary workshops utilizing locally grown ingredients. "Early French education sorts out students according to aptitude. When one starts the culinary arts, usually by age fourteen, emphasis is on technical training. This technical training, which of course includes all the basics which the French codified as modern cuisine, is deeply influenced by locale and local culinary tradition. Even the great chefs in Paris always cling to some degree to their regional roots and respect for local products."

The farmers' markets spill over in France, when on one day of the week, traffic is closed down and everything stops for the market. This country, among some other European countries, is known for subsidizing its farmers, almost as though it sees them as its open-space park rangers that need to be paid for their service in addition to the little they receive for supplying food and fiber (knowing these basic products must, in general, stay priced relatively low.) France is also known for helping farmers become agritourism destinations, where visitors can travel from castle to quaint village with French

A room in a centuries old French farmhouse is converted to a B&B guests' dining area for overnight agritourists who stay at the farm.

countryside in between, and stop over at a generational farm for an experience of French farm life, food, and rest.

"In Europe," says Leslie Zenz, former manager, Small Farm and Direct Marketing Program at Washington State's Department of Agriculture, "hundreds of Italian and French agriturismos offer guests the chance to participate in, for instance, late-summer grape harvesting, crushing, and winemaking. All the European agriturismos are subsidized by the government. They receive stipends and assistance with capital costs because Europeans see it as a way to help preserve the remnants of a long agricultural history. I don't see the U.S. policy adopting that model in any widespread fashion any time soon."

What about the people who own and operate the successful French agritourism farms? The ones we met are kind, yet quite happy to be who they are rather than what foreign tourists might want them to be. At a generational farm located in southern France, the owner offered to let us see her elderly father's private museum of cave people's artifacts. What time was best for us to visit, she asked. What? Noon? Heavens, no! The French stop all activity

between noon and two to have their midday meal. No activity other than food services goes on at that time. They don't change to cater to tourists for anything.

Further on in France, I sipped my fresh raw cow's milk, noting the small lumps of buttery cream that floated on the top while dining in the lovely, stony farmhouse of a working dairy in the fertile Louvre Valley. We shared the dining area with another couple from England, so among us, English was being spoken. The attentive farmer brought out our dishes as his wife had them prepared. But he wouldn't attempt a single word in English. We know (we know) he knew at least a little. But we were in France, and that's apparently what we were going to get as far as he was concerned, everything French. We loved the feeling of adventure and challenge this brought. We were just learning the French language ourselves, and wanted to experience the real thing. We didn't want a French satellite homogenization of a luxury, multinational hotel with a little Herbs de Provence sprinkled on our food so we could call it "French."

One confidential European interviewee admitted, "It isn't perfect over here." There is concern

This farm dog loves to wander into the dining area to mingle with the French farm B&B guests. In France, dogs and cats of guests and owners are allowed into the restaurants.

that "boutique" farms may replace authentic agriculture to receive funds for the hospitality business. Numerous, frustrated Parisians and one citizen where we stayed in an historic, apple cider farm in Normandy said their country was very slow to transition to organic methods, which they felt had the potential to benefit the countryside and agritourism even more. Traditional farming is there in France, yes. But since, little by little, pesticides and refined white flour had crept their way into their traditional farms and baguettes, change back out of those "traditions" comes very slowly there. We can learn, perhaps, to watch out for the type of negative change that creeps its way in, little by little, hoping we won't notice until one day, we think "it's always been this way." We can be alert, rather than complacent, to the problem of tradition that comes from the fear of any change at all, as well as its polarity, changing for the thrill of something new. We may realize, for example, that a sustainable new future may be something valuable and worth holding onto.

When government values and supports its local farmers, there is great potential in the making, but it still needs to be done correctly. For example, some have seen government funds offered to farmers for on-farm hospitality, but when the farming families know nothing about the hospitality business, a disaster is waiting to happen.

Compared to the USA so far, however, the French give us a glimpse, and at least partially demonstrate, the potential of what could happen when the government supports local, traditional but progressive farming and its country's own flavor, style and tradition. From this, we could see how grand it could be if the ongoing advancement of sustainable practices and hospitality training were also ensured. It's a direction many of us would like to reach for… a government that's aware of how supporting its farmers as dignified and valuable business people mutually benefits the entire country, and where there's great pride in tourism and preserving history.

But even though government funding can be of great assistance, some feel it may be best to consider it a supplement towards eventual, complete self-sustainability. Others feel it's a waste of time to wait around for it at all.

Tourists who take part in Cretes Culinary Sanctuaries can witness people like Panayiotis Moldovanidis, shown here collecting wild dandelion and wild sorrel in eastern Crete.
Photo by Nikki Rose

EPILOGUE

Crete's Culinary Sanctuaries helps tourists meet people like Katerina Tzanaki in a manner that helps them progress while sustaining their traditional livelihoods, shown here with her own artisan cheeses. She and her husband are farmers, artisan cheesemakers, and bread bakers in an extremely remote village in Eastern Crete. She's holding a block of her precious "Anthotiros." "Anthos" means flower -- referring here to the aroma of the cheese as one would refer to a flower's aroma. "Tiros" means cheese in ancient Greek. Tourists learn that traditional Cretans still use many ancient Greek words.
Photo by Nikki Rose

The Island of Crete

On the island of Crete, native wild edible plants including oregano and artichoke are harvested gently, and small-scale, sustenance farming is being practiced as it has been for more than four thousand years. Honey is spun as it has been for centuries, and thanks to someone named Nikki Rose, eco-tourism workshops are set up for practitioners (the farmers, bakers, etc.) to offer tailored seminars for small groups. Seasonal programs include the subjects of healthy cuisine, history, nature and agriculture, such as the production of organic olive oil and cheese. Visitors have a rare opportunity to experience history and tradition from an insider's viewpoint. One must actually visit to discover first hand the folklore, music, and methods of culture and farming that reveal a deep, rich understanding of the earth and human society. Today, according to Nikki, modern researchers are using these traditional models to provide solutions to today's issues relating to industrial farming and toxic pesticides.

Nikki Rose is a multiple award-winning Greek-American professional chef, writer and one whom some describe as a visionary and socio-cultural activist. In 1995, she began organizing cultural-culinary seminars in Washington D.C. to generate consumer interest in the work of culinary artisans, small-scale farmers and producers. Prominent chefs in the diplomatic and private sectors joined in to further the cause. Then she moved to Crete in 1998 to organize small-scale cultural tours. Here, she founded Crete's Culinary Sanctuaries (CCS) in 1998 to support people actively working to preserve cultural heritage and the environment, including organic farmers, chefs, eco-lodge owners, holistic health practitioners, and nature conservationists. She had been witnessing Crete's rural population's rapid decline along with its traditional trades and natural beauty. This is when she shifted focus to preservation on all fronts—rural communities, trades and nature, bringing them all together in a network of mutual support. CCS programs now promote the work of local farmers, culinary artisans and related groups, which helps to supplement and sustain a variety of preservation projects.

Traditional Crete, however, faces extinction partially from tourism itself—big business, generic, resort-type tourism—and even from the very thing that authentic agritourism would ordinarily want to attract: the media. The fork in the road has appeared, and Nikki is dedicated to steering agritourism in Crete in the right direction. She knows that agritourism has huge potential, that "the big guys" and the small ones can live side-by-side for mutual benefit, and she also knows the difference between the authentic and superficial.

As far as agritourism's potential in Crete and worldwide, she has many good things to say: "First, just by having other members of the community (CCS and fellow farmers in our case) acknowledge their great work helps," Nikki says. "Farming is hard work that is rarely lucrative, and in the modern sense of the word, all encouragement helps. It's also not a glamorous profession, and the children of farmers around the world now (usually) have access to international, glamour news, real or imagined, and they don't want to carry on the family farm. Agritourism can bring family farms back into the mainstream and encourage children to sustain the business and see it from a different perspective: They have the opportunity to simultaneously strengthen their community, help preserve their cultural heritage and protect their environment. If the children in farming communities who already have the expertise to carry on the business," Nikki continues, "vanish into the modern world, the farming communities vanish with them. Sound, long-term agritourism strategies may help them to sustain their family farms, and make a decent living, which is what all farmers deserve." Nikki also sees how agritourism can form a strong network of support for farmers. "Since small-scale organic farming is rarely supported by governmental officials as it should be," she says, "many farmers are working on their own and feel alienated. There's big campaign money to be found in the industrial farming sector (machinery, pesticides, seed companies). Simplistically, yet realistically, small-scale farmers will not be buying a $10,000 table at an election fundraiser. Their opinions about sustainable development are rarely heard. We have seen how damaging political alliances with industrial agricultural companies have been around the world, in terms of irreversibly damaging environmental impacts, the health of our soil, the safety of our food, the indigenous flora and fauna, and the dissolution

Crete citizen Panayiotis Moldovanidis shows how to collect wild asparagus as has been done for centuries. It is very small and delicious, and grows like a vine among sticker bushes.
Photo by Nikki Rose

Nikki Rose on a mission to collect wild greens in Crete She is the founder of Crete's Culinary Sancutaries which was selected as a concrete case of good, sustainable practices worldwide by the World Tourism Forum for Peace and Sustainability.
Photo by Panayiotis Moldovanidis

of farming communities. Agritourism can link farmers directly with consumers and re-strengthen that necessary bond. Agritourism can break the damaging barriers created by politics and industrial farming, between small-scale farmers and consumers."

But problems arise when local, luxury resorts attract tourists by using the charm of rural life without working with the farmers themselves, making sure part of the revenue somehow goes back to them and making sure their lifestyle is maintained, if not upgraded, for years to come. Farmers can't just stop their work any time they're asked to show their crops, for free, to a group of customers sent over by a resort, or a group of filmmakers sent over by the government to hopefully draw more tourist dollars to the country. As the cost of living rises while the income of small-scale farmers and artisan food producers remains low, continued exodus from rural communities is inevitable.

And as professional writers ourselves who understand the value of true journalism, Nikki and I also have concerns about "staged media," such as fuzzy, feel-good "documentaries" set up, once again, to attract throngs of visitors to the area to witness 4,000 year old farming history, with the resulting tourist revenue going to the resorts, and nothing set up ahead of time to make sure the time away from work to shoot the "documentaries" will pay off for the farmers.

"In our work," Nikki says, "we find out what farmers are doing now, and their future plans. We bring in anyone in the community that may be able to help make their visions a reality. We also discuss the options of how to present their work to visitors... this is a very important factor. People in farming communities tend to think that what they do is not interesting... but it is to the outside world. So, each farm needs to have a purpose... a reason for people to want to visit. We bring out the best in what they do, and convey that to visitors.

"When people visit farmers and discover their wonderful products and great passion for what they do, agritourism is working. Farmers also need to consider branding a few of their products to generate income beyond the revenues they might earn by simply charging an entrance fee to their farms. This is challenging because of the daunting rules for organic certification and production. In Crete, there are many farmers we work with who still sell their food in bulk only. What we have to do is create more activities such as cooking demonstrations that generate more income for them. There's also liability issues that small-scale farmers cannot afford to address (for numerous reasons) when they open up

Though water is currently scarce, the local community in Cape Town, South Africa, has found ways to build eco-cottages with very low water use.
Photo courtesy Shaster Foundation

their farms to visitors. Insurance policies can be daunting and vague... and consequently limit the visitor activities they can plan on their farms.

"We cannot place the burden of preserving tradition up to a few people in our world, at their expense alone," Nikki continues. "If we think tradition should be offered to us for free or at a discount while supplies last, we are part of the serious problem. The people we meet during our travels are not obligated to treat us to a single peanut, let alone a feast. They are not our servants or our free entertainment—no matter who we are—public officials, entrepreneurs, journalists, researchers or travelers. Tradition is not ours to take. It is ours to preserve. If we want to be part of the solution, we need to be the solution. Ethical businesses cannot survive without ethical patrons."

Nikki is a wealth of experience whose insights go much further than the room available to her in this epilogue. She is a busy person with many people vying for her time. For further information, see what she writes on her website, Crete's Culinary Sanctuaries (see Resources), to learn what she has to say about furthering the cause of authentic eco-tourism.

So, once again, as Nikki demonstrates, seek out the correct audience for your agritourism, those ethical patrons, which will form that beneficial alliance, bring strength in numbers, bring support from diverse areas and protection from misguided or uninformed government and big business. Select media carefully if they become overwhelming or suspicious. Nikki has worked with some wonderful media. As a writer herself, she knows how much work must go into the discovery and conveyance of a deep story that readers will enjoy. But she's very careful about choosing the media with which she shares her precious time.

Cape Town, South Africa

South Africa, the land of ruby-hued rooibos tea, native, scented geraniums, and where the original ancestors of our sweet melons still grow wild. I had hooked up with a unique group of African citizens several years ago who knew incredible knowledge of ancient African and regional organic farming, eco-tourism's potential, and the natural healing benefits it could endow upon many layers of society. Today, I was traveling there to visit with tribal elders to listen to a fascinating plan for an organic farm and eco-tourism center to help Africa help itself. Our plane landed in Cape Town after a long, coach-class 12-hour flight from London. I didn't even know what day it was as we staggered to pass through

South African child in food garden. South Africans are using eco- and agritourism to fund their sustainable projects that are helping lift themselves out of despair.
Photo courtesy Shaster Foundation

Clearing land for one of the food gardens in the shacktown built in a rubbish dump.
Photo courtesy Shaster Foundation

The first handsome eco-cottage nears completion.
Photo courtesy Shaster Foundation

customs. The dark, tall uniformed man with a strong Afrikaans accent took my passport, checked paperwork, smiled warmly and said, "Welcome to South Africa… And happy birthday." He knew better than I that little things are important, and that at least for me, this was a special personal holiday.

Africa is a place of conflict and destitution, yes, but of warmth, hope, immense intelligence, and huge potential as well. And as monetary aid, often superficial, continues to pour in while problems continue to mount, in a few pocket places there are Africans who insist that it is within the culture itself to rise above devastation. They refuse aid from offers for GMO (Genetically Modified Organism) seeds and turn to organics, their own open-pollinated seed supply, and permaculture instead. They are even cautious about vaccines that don't address the underlying problem and lead to even more dependency and illness in the future. They ask for a hand up "as long as your own are taken care of, first," but not for an external bandage that makes heroes out of the wealthy from other countries while "victimitis" and deeper problems within their borders continue to fester and form into yet more problems down the road.

During this trip and subsequent interactions with my friends from South Africa, we formed an alliance to help these small and scattered groups of people grow, spread, and become jewels first to themselves, and then to the world, while large amounts of aid money poured into other venues, refusing to siphon any their way. I learned of the healing potential of milk from exotic African animals whose names I could barely pronounce. Of healing herbs and a multi-tiered agritourism plan that would wrap around a native African organic farm, healing and arts center. My main correspondence now in Cape Town is Di Womersley, a native African, and her non-profit in South Africa is the Shaster Foundation, which is now an "adopted sister" to my own USA non-profit. She is assisted in South Africa by her partners, Makhosi Athobile, a highly respected traditional healer (called a sangoma); Stanley Ndlovu, one of South Africa's most talented sculptors eager to teach art to young Africans; Geoff Ainslie, an innovative businessman; and Dizu "Zungula" Plaatjies, an internationally known South African folk musician.

From the USA, we sent open-pollinated seeds and books on small-scale farming, including *Micro Eco-Farming: Prospering from Backyard to Small Acreage in Partnership with the Earth* by this author, and *The All New Square Foot Gardening* by Mel Bartholomew. The seeds were carefully shared out and grown in private gardens to proliferate the open-pollinated seed supply. The people there drove out to the farmers to warn them against GMOs. As time progressed, hopes to secure some promised land for this center fell through (again, being targeted instead to more conventional aid). So what we had was a growing number of seeds and sustainable farming knowledge, the need for a healing and learning center, an eco- and agritourism idea, and a township built upon an old garbage dump. Here, makeshift shacks of tin and old wood are crammed together. Neither running water nor electricity exists. As my group back home in the USA and others continued to send more open-pollinated seeds and books, other smaller donations found their way there, and our friends in Africa built their first mini-sustainable farm—a tiny food garden—along with a worm composting toilet, right in the midst of the shack town near a "crèche," which is what South Africans call a day care center for young children, and in this case, including orphans.

And then something unusual happened. Here's a segment from a letter from Di Womersley sent out to supporters:

(See Sidebar, "Letter from Di #1")

As time passed, others began to discover the Shaster Foundation and donate ideas and resources. A healing center for the township has been built. With the help of their growing supply of open-pollinated seeds and new knowledge from the permaculturists and *Square Foot Gardening,* a plan has been set in motion for an eco-agritourism project: they are building a hikers' hostel and eco-cottages surrounded by organic food crops. This is taking place right in the midst of the rubbish shacktown, both to help create authentic homes in the area, as well as to fund the restoration of the township via agritourism and eco-tourism.

(See Sidebar "Letter from Di #2")

The first, actual eco-cottage became available close to the planned time. For every 200 paying guests, another real cottage can be built for a family living in a make-shift shack, with the goal of replacing every shack with a self-built eco-cottage. The cottages are made from locally sustainable materials, with composting toilets, wind power and organic gardens, and built with help from the people themselves, including unemployed youth. From within, this shack town is slowly being transformed to a self-sustaining eco-village.

We still hope for land for a separate, large self-sustaining eco-center to accelerate this progress within the current township project, and to be able to research, host, teach and serve many more from the surrounding area to do the same. A research and sustainable demonstration farm could train the growing numbers of eager new African sustainable farmers, supply and train more orphans and families in need to help themselves out of poverty, and supply on-farm, authentic African agritourism and an eatery to draw revenue. It could also research promising new information linking areas of specific soil micronutrient depletion to various illnesses, and possibly serve as a study area for the Terra Preta project, in which soil discovered in a few places in Africa, South America and others is reputed to retain its fertility and increase crop yields to a far greater magnitude than what has so far been attained by any other method. (See AcresUSA.com for more information on this project. They give ongoing reports and you can search their archives to find past information). Volunteer teaching and other help with the orphans would be a destination activity in itself for visitors from around the world.

Our group in Cape Town is as an example to those farmers everywhere who may have grand goals, but think they don't have the means to begin, or that they are helpless without large sums of money from big organizations. The lessons shared by this courageous group of South Africans might be to start where you are, keep moving ahead one step at a time, and notice how support can seem to come out of nowhere. Once again, wherever you begin,

Letter from Di #1

– Letter from Di Womersley, director the Shaster Foundation, a non-profit foundation in South Africa, to supporters

Greetings:

If anyone is ever feeling jaded about the state of humanity and despairs of hearing any good news, then just connect with us and we will fill you in on some of the amazing things that are happening here in Cape Town.

Our project at the Ndlovu Crèche (now renamed as the Ndlovu Centre because it is far more than just a crèche) in Khayelitsha is moving forward, upward and outward at a terrific pace.

Let's begin with the garden. Permacore (Permaculture Foundation of the Western Cape) came out in force and worked side by side with the street committee to clear the plot, sheet mulch, manure and plant a huge variety of vegetables, which are being harvested regularly, and we have kept up a supply of manure/compost until there is a compost heap ready on site. Neighbours are also growing veggies, and seeds are being saved. Thanks to a wonderful lady in the USA, Barbara Adams and the World Grace Foundation, who have sent so many seeds over, organic pumpkins and mielies are amongst some of the veggies just planted. Just the sight of growing plants lifts the whole ambience of the area. We also have a 5,000-litre water tank thanks to Johan van Vuuren, and guttering to fit onto the building to catch the rain run-off. There is no running water nearby—it all has to be fetched by hand. So the kids now wash their hands over a vegetable before meal times to save water. Hopefully we still will have enough rain coming to fill the tank before the heat of summer.

Set amongst the spinach plants is a thing marvelous to behold! A bright, blue toilet, in which thousands of earthworms are labouring intensely to make the compost that will feed the veggie garden, which will eventually come back to feed them… The only toilets in the area are foul-smelling "long drops"—or the bushes. There have been many incidents of women being attacked when they go into the bushes to relieve themselves—how many of us take something as simple as a toilet for granted? We hope to put in toilets for the rest of the houses in the street soon.

Downstairs in the kitchen sits the new stove and fridge which are hard at work every day. In the crèche area the children now have carpets on the floor as well as mattresses and blankets for nap time. They have an abundance of toys (thanks to the Beehive Montessori pre-school) which keep them occupied, and by the beginning of next year one of the caregivers will be a properly trained, pre-school

continued...

Letter from Di #1 ...cont'd

teacher (thanks to Kyoko who originally introduced us to the crèche, and who was the instigator of it all). There will be a grand party to declare the crèche officially open some time in January—anyone interested in the project is most welcome to attend.

After the little ones go home at 4 p.m., the youth group of 14 girls aged between seven and 17 years work hard at practicing traditional songs and dances for on average three hours every day. They are very proficient, and we are working on getting the materials to make good costumes for them. A new drum makes practice a great deal easier!" Many thanks to Dizu Plaatjies for this.

In the mornings this space (the crèche that took care of the children during the day) will be used by the Ilima Bafazi womens' craftwork project – we have 2 large floor looms and several small table looms – some donated and some sponsored by JDI. We have applied for funding to start teaching weaving to 14 women who are either HIV positive or TB patients and are all unemployed. Their children will be taken care of in the crèche while they are working upstairs. We have been given three sewing machines, and large bundles of thread, wool, string, etc. Within six months we hope to have a thriving, small industry here once sufficient funds have been raised for training. "Already a women's group is making traditional meals for visitors who would like to have a safe and authentic township experience. This summer they will also be able to enjoy some traditional singing and dancing, and soon we hope to have a small shop for them to buy items made in the craft project. A visit to this dynamic and friendly community has inspired many people to get involved in some way, and we hope more and more people will be visiting from overseas.

We plan to be able to give every family that cares for orphans: food-gardening training, seeds, manure and tools to start growing food if they are unemployed.

I would also like to share something heartwarming to anyone who has ever done volunteer work in economically deprived communities. The street committee at the centre has collected together R1,000 which they have said is the community contribution (as well as doing all the building work) towards making their crèche very "professional" for the sake of the children. Nowhere else I have worked has anything like this happened. It just makes me even more determined to make this a resounding success.

Love in Abundance,

Di

Letter from Di #2

Greetings dear Barb!

I hope this year brings wealth, health, peace and prosperity to you and all your loved ones.

2007 has started with a bang for us... some wonderful donations from around the world, including from a well-known actor, have enabled us to begin building 2 eco-cottages, finish the clinic and plan an educational centre and backpacker hostel. Our little project is taking wings and flying. I truly believe that the goodwill and kindness of people like yourself and all your supporters adds to the great flow of positivity that this world needs so desperately.

By the end of February we will have the cottages built and gardens (square foot style!!) planted.

Warm African love from a hot and sunny Cape Town

Di

Note: Since this letter, Di has written that visitors to the health center, which uses only natural medicines, more than tripled.

---

aim for the right audience, those who want to see you just as you are, and help you grow from there as they support you with their paid visits.

Concluding thoughts

From ancient, tribal communities worldwide to abundant times in early America, farms and farmers, simply by nature of how human society naturally functioned, weren't isolated from their non-farming community and its commerce structure, marketing needs, new discoveries, and social network. In our post modern era, mending that isolation and restoring the needed connection now seems to need formal names, such as agritourism, local food revival, and I've even heard the term, "contact era agriculture." But the farmer-community connection started out very natural and informal and has always been an element of successful farming.

This book was written as a contribution to the preservation of farming. It's never been my desire to help the farming industry stay stagnant and cling to a definition of "real farming" that has proven itself disastrous. This includes "farming" defined as an isolated segment away from the social structure as a crop planter or herd steward, with little voice or interaction with the non-farming community or how the world is progressing and changing. Such a set-up is dependent on payment from business owned by outsiders, owned by those who know nothing about the business of farming or don't want to know. The definition of what it means to farm and be a farmer is rewriting itself. Those who want only to work the fields may be better and safer as employees of high-integrity, regionally-based businesses rather than business owners themselves. As the Leopold Center for Sustainable Agriculture states, for years, farmers have been told only how to get more yield out of certain seeds, rather than

Carrots in South Africa going to bloom to create next year's seed.
Photo courtesy Shaster Foundation

helping them relearn business and marketing and plug into the world economy, where they can choose for themselves which seeds to plant and whether higher yield resulting in oversupply is even necessary.

Agritourism is one of many business tools that can help maintain and uplift the lifestyle of farmers and strengthen the foundation of a sustainable, local farm economy. Like everything else, it isn't flawless, nor is it for everyone. But for those interested, it is available in degrees of intensity, from a sign for an honor-system flower stand for passersby, to inviting guests from around the world to stay overnight. Like any tool, its effectiveness in contributing to a sustainable economy and restoring local farms depends on the agenda and awareness of those using it. In the future, it is my hope that agritourism remain out of the hands of those who turn blades into swords and remain owned and guided by those who use blades as plows, hoes, and trowels. May every country and continent rise into an abundant and self-sustaining nation operated by the voices of aware and intelligent locals. And may your own standard of living rise as your hay fields remain free from slickers who finally learn not to trample them, and your local school children one day know that chocolate milk doesn't come out of brown goats. ❧

For further information on agritourism, visit this book's companion website:

www.NewAgriTourism.com

Resources

Associations, Resources

(The) New Agritourism Companion Website. Color photos tours of successful agritourism farms and other helpful resources

www.NewAgritourism.com

100-Mile Diet. The official homepage of Canadian "locavores" Alisa Smith and James MacKinnon, identifies your 100-mile radius and publishes success stories from all over the nation.

www.100milediet.org

Acres USA. "A voice for eco-agriculture," including monthly publication, conferences, books.

www.acresusa.com

American Independent Business Alliance, The. Defends community character by helping sustain independent businesses, raising awareness of their needs and otherwise unknown struggles, and networking to strengthen communities.

www.amiba.net

Bangladesh Rural Advancement Committee

www.brac.net

Chefs Collaborative. National network providing members with tools for running economically healthy and sustainable food service businesses.

www.chefscollaborative.org

Community-Wealth.Org. Community-based wealth strategies, policies, models, and innovations. The organization connects community corporations, co-ops and nonprofits to create a network of support and participation.

www.Community-Wealth.org

Crete's Culinary Sanctuaries. Nikki Rose's website, mentioned in Epilogue of this book, dedicated to furthering the cause of authentic eco-tourism.

www.cookingincrete.com

CSA-L E-mail list and discussion group on Community Supported Agriculture (CSA).

www.prairienet.org/pcsa/CSA-L

Ecological Options Network, The. Gathering site for activists, innovators, and information from around the globe for ideas about existing, sustainable alternatives.

www.eon3.net

Envirolink. Online environmental community, a nonprofit providing access to thousands of online environmental resources from agriculture to wildlife. Check their resource links: sustainable business and sustainable living.

www.envirolink.org

Farm to College offers a look at all the Farm to College programs in the United States, with statistics about percentages of local foods purchased, finances devoted to the program, etc.

www.farmtocollege.org

Farm to School helps establish local food programs in school cafeterias, and educates students about the benefits of eating local food. Includes a map of current U.S. programs.

www.farmtoschool.org

Farm-Based Education Association

www.farmbasededucation.org

Farmer-Chef Connection, The. Allows farmers and restaurant-owners to "speed-date," matching up everyone's best interests and forming mutually beneficial business arrangements. These partnerships can help increase the availability of fresh, locally grown products used in mainstream restaurants.

www.ecotrust.org/foodfarms/farmerchef.html

Grassroots Economic Organizing offers a newsletter on grassroots organizing to promote networking, build and finance worker-owned, democratically run, community based, ecologically sustainable enterprises.

www.geo.coop

Leopold Center for Sustainable Agriculture, The. Research group in Iowa that investigates the dangers of current farming practices, educates about its very innovative findings, and proposes great alternatives.

www.leopold.iastate.edu

Local Legacies. List of local festivals and other community events celebrating the unique heritage of each U.S. state. Includes links to each one's website.

www.loc.gov/folklife/roots/

Market Farming E-mail list and discussion group plus helpful articles on market farming.

http://lists.ibiblio.org/mailman/list info/market-farming.

Networking Association for Farm Direct Marketing and Agritourism (NAFDMA). Trade association dedicated to nurturing the farm direct marketing industry.

www.nafdma.com

National Trust for Historic Preservation, The. Information on how to protect and restore the residential and business areas of towns, preserving local history and unique character. Resources are categorized by state and by topics of interest. The home page is

www.nationaltrust.org/
community/

New Farm: Regenerative Agriculture Worldwide, The. Possibly one of the best resources available on worldwide sustainable agriculture, including occasional information on agritourism.

www.newfarm.org

Plant A Row For the Hungry

www.gardenwriters.org/Par/index.
html

Relocalization Network, The. An initiative of the Post Carbon Institute, supports groups worldwide dedicated to localizing their food and energy economies.

www.relocalize.net

Slow Food International. Organization devoted to protecting regional foods from homogenization worldwide, with branches in many countries. Sustainable farm and gardening resources, including a magazine with occasional articles on bringing customers to the farm.

www.slowfood.com

Social Investment Foundation Forum, The. A huge bank of resources, contacts, and media releases about sustainable living.

www.socialinvest.org

Books, Periodicals, Publications

Acres USA. Monthly publication.

www.acresusa.com

Bringing the Food Economy Home, by Helena Norberg-Hodge, Todd Merrifield, and Steven Gorelick. Discuss the localization of our food economies and how this is a "solution-multiplier" that will reduce the negative impacts of globalization. Zed Books, 2002.

Eat Here: Reclaiming Homegrown Pleasures in a Global Supermarket, by Worldwatch senior researcher Brian Halweil. Discusses the private and global benefits of eating locally, and the problems of foods that must travel many miles. W. W. Norton & Company, 2004.

Eco-Foods Guide, The: What's Good for the Earth Is Good for You! by Cynthia Barstow, New Society Publishers 2002.

ECOpreneuring: Putting Purpose and the Planet before Profits, by Lisa Kivirist & John Ivanko, New Society Publishers (available July 2008). "Part small business manifesto, part personal finance primer, ECOpreneuring dives into what being green means in the 21st Century and why small, ecopreneurial businesses are essential to recreating a better world." New book from the authors of *Rural Rennaissance.*

Farming Alternatives: A Guide to Evaluating the Feasibility of New Farm Based Enterprises. Grudens-Shuck, N. and J. Green. 1991. Farming Alternatives Program, Cornell University, Ithaca, NY. 88 p. This publication uses a step-by-step process to assess goals, resources, markets, etc. Includes worksheets.

From Mondragon to America: Experiments in Community Development, by Greg MacLeod, describes the success of the community-owned town of Mondragon, Spain, including how to adapt and replicate its business and social experiment in other communities. University College of Cape Breton Press, 1997.

Full Moon Feast: Food and the Hunger for Connection, by Jessica Prentice, Chelsea Green Publishing 2006.

Grub: Ideas for an Urban Organic Kitchen, by Anna Lappé and Bryant Terry, is a wonderful collection of recipes and stories that illustrate the importance of an organic lifestyle as personal health as well as environmental health and societal justice. Tarcher, 2006.

Hobby Farms Magazine and Hobby Farms Home. Rural living info, resources.
  www.Hobbyfarms.com

In Defense of Food: An Eater's Manifesto, by Michael Pollan. Penguin Press 2008.

New World Publishing: Publisher of this book, as well as other books on small/sustainable agriculture, including Micro Eco-Farming (by Barbara Berst Adams), The New Farmers' Market, and Sell What You Sow!
  www.nwpub.net

Primer for Selecting New Enterprises for Your Farm, A: Woods, Tom and Steve Isaacs. 2000.. Cooperative Extension Service. University of Kentucky. Agricultural Economics - Extension No. 00-13. 28 p. Covers profitability, resources, information, marketing, enthusiasm, and risk. Has many useful worksheets from which accurate information can be generated to guide your decision making.
  http://www.uky.edu/Ag/AgEcon/
  pubs/ext_aec/ext2000-13.pdf

Primer on Community Food Systems, A: Publication of Cornell University offering solutions for community food systems.
  foodsys.cce.cornell.edu/
  primer.html

Small-Mart Revolution, The, by Michael Shuman, reveals detailed information on how small businesses are triumphing over multinational chains. Berrett-Koehler Publishers, 2006.

Way We Eat, The: Why Our Food Choices Matter, by Peter Singer, Rodale Books 2006.

Specialty Seeds, Seed Saving

Abundant Life Seeds
  www.abundantlifeseeds.com

Association Kokopelli
  www.organicseedsonline.com

Baker Creek Heirloom Seeds
  www.rareseeds.com

Heirloom Acres Seeds
  www.heirloomacresseeds.com

Johnny's Selected Seeds
  www.johnnyseeds.com

Marianna's Heirloom Seeds
  www.mariseeds.com

Native Seeds Search
  www.nativeseeds.org

Nichols Garden Nursery
  www.nicholsgardennursery.com

Pinetree Garden Seeds
  www.superseeds.com

Seed Savers Exchange
  www.seedsavers.org

Seeds from Italy
  www.growitalian.com

Seeds of Change
  www.seedsofchange.com

Select Seeds - Antique Flowers
  www.selectseeds.com

Southern Exposure Seed Exchange
  www.southernexposure.com

Territorial Seed Company
  www.territorialseed.com

Turtle Tree Seed
  www.turtletreeseed.org

Underwood Gardens
  http://underwoodgardens.com

Victory Seed Company
  www.victoryseeds.com

West Coast Seeds
  www.westcoastseeds.com

Index

Thank you for purchasing *The New Agritourism!* If you like it, please tell others! Quantity order discounts: See *Orders* below.

Micro Eco-Farming: Prospering from Backyard to Small-Acreage in Partnership with the Earth, by Barbara Berst Adams. Micro eco-farmers across the nation are profiting from backyard to small acreages. *Micro Eco-Farming* shows what you can grow, innovative farming methods you can use, and how to market products in creative, new ways. 6 x 9, 176 pps., $16.95.

Sell What You Sow! The Grower's Guide to Successful Produce Marketing, by Eric Gibson. This book was described by *Small Farmer's Journal* as "far-and-away the outstanding farm produce marketing text." 8 1/2 x 11, 304 pps. $28.95.

The New Farmers' Market: Farm-Fresh Ideas for Producers, Managers & Communities, by Vance Corum, Marcie Rosenzweig & Eric Gibson. Definitive book on farmers' markets for farmers, market managers or city planners. 8 x 10, 272 pps. $26.95.

Free Downloads: The Hot 50 Marketing Tips, Top Trends in Farmers' Markets, Benefits of Farmers' Markets for Farmers, Customers & Community, Selling to Ethnic Groups, Getting Top Dollar for What You Sell, and more! Online at www.nwpub.net.

Single orders: Call 888-281-5170 or online at www.nwpub.net.
Quantity orders: Call 530-823-3886.
New World Publishing, 11543 Quartz Dr. #1
Auburn, CA 95602. Ph. & Fax: 530-823-3886
nwp@nwpub.net • www.nwpub.net

About the Author

Barbara Berst Adams organized numerous agritourism events on her farm while her children were growing up there. These included apple pressings, autumn harvest and pumpkin festivals, petting zoos, pony rides, hay rides, historical re-enactments, and various workshops and retreats. She has traveled to South Africa, France, Holland, England, Switzerland, Mexico, Canada and across the USA to witness how agritourism is benefiting farms, strengthening local communities and expanding humanity's eco-awareness. She and her husband, Kipp Davis, are now restoring Island Meadow Farm, the original farm her children grew up on. She invites readers to visit this book's companion website: www.NewAgriTourism.com. Here you will find updates on the agritourism frontline, more ways for farms to connect beneficially to their local communities and the world, plus stories and links to the agritourism farms that surround her Island Meadow Farm in the emerald islands and fertile valleys of Northwest Washington State.

Barbara also is the author of *Micro Eco-Farming: Prospering from Backyard to Small-Acreage in Partnership with the Earth,* published by New World Publishing.